[redacted]

MATHEMATICAL KNOWLEDGE

Mathematics Education Library

VOLUME 10

The titles published in this series are listed at the end of this volume.

MATHEMATICAL KNOWLEDGE: ITS GROWTH THROUGH TEACHING

Edited by

ALAN J. BISHOP, STIEG MELLIN-OLSEN

and

JOOP VAN DORMOLEN

for the BACOMET group

KLUWER ACADEMIC PUBLISHERS
DORDRECHT / BOSTON / LONDON

Library of Congress Cataloging-in-Publication Data

Mathematical knowledge : its growth through teaching / by Alan J. Bishop, Stieg Mellin-Olsen, and Joop van Dormolen (eds.).
p. cm. -- (Mathematics education library Volume 10)
Includes index.
ISBN 0-7923-1344-5 (acid free paper)
1. Mathematics--Study and teaching. I. Bishop, Alan J. II. Mellin-Olsen, Stieg, 1939- III. Dormolen, J. van. IV. Series.
QA11.M3753 1991
510'.71--dc20 91-23234

ISBN 0-7923-1344-5

Published by Kluwer Academic Publishers,
P.O. Box 17, 3300 AA Dordrecht, The Netherlands.

Kluwer Academic Publishers incorporates
the publishing programmes of
D. Reidel, Martinus Nijhoff, Dr W. Junk and MTP Press.

Sold and distributed in the U.S.A. and Canada
by Kluwer Academic Publishers,
101 Philip Drive, Norwell, MA 02061, U.S.A.

In all other countries, sold and distributed
by Kluwer Academic Publishers Group,
P.O. Box 322, 3300 AH Dordrecht, The Netherlands.

Printed on acid-free paper

Printed in the Netherlands

CONTENTS

INTRODUCTION

In the first BACOMET volume different perspectives on issues concerning teacher education in mathematics were presented (B. Christiansen, A.G. Howson and M. Otte, *Perspectives on Mathematics Education*, Reidel, Dordrecht, 1986). Underlying all of them was the fundamental problem area of the relationships between mathematical knowledge and the teaching and learning processes. The subsequent project BACOMET 2, whose outcomes are presented in this book, continued this work, especially by focusing on the genesis of mathematical knowledge in the classroom. The book developed over the period 1985–9 through several meetings, much discussion and considerable writing and redrafting.

Our major concern was to try to analyse what we considered to be the most significant aspects of the relationships in order to enable mathematics educators to be better able to handle the kinds of complex issues facing all mathematics educators as we approach the end of the twentieth century. With access to mathematics education widening all the time, with a multitude of new materials and resources being available each year, with complex cultural and social interactions creating a fluctuating context of education, with all manner of technology becoming more and more significant, and with both informal education (through media of different kinds) and non-formal education (courses of training etc.) growing apace, the nature of *formal* mathematical education is increasingly needing analysis. Formal, institutionalised, mandatory mathematics education is what schools provide and whilst practices vary between schools, between regions, and between countries, there are nevertheless significant and meaningful similarities which can be documented and analysed.

The BACOMET group presents here its work on the genesis of mathematics knowledge in formal education. This topic presents an exciting field of research for those entering into it. Not only is there the rich development in the research disciplines which support the science of the didactics of mathematics, but there is also the courageous thinking of several innovators in the field of mathematics education and the emergency of various epistemologies underlying learning which provide the mathematics educator with new perspectives for research strategies, new tools for innovative work, and new possibilities for promoting the learning of mathematics. The BACOMET group contributes to such developments by attacking the general issue of the genesis of mathematical knowledge through the different experiences, perspectives and scientific traditions which its various members represent.

Alan J. Bishop et al. (eds.), Mathematical Knowledge: Its Growth Through Teaching, 1–9.

FIVE MATHEMATICS CLASSROOMS

In order to illustrate the sort of table we shall lay before the reader we first present five cases of mathematics learning taken from five rather different mathematics classrooms. The first example, quoted in the BACOMET 1 volume, concerns an elementary school class beginning to come to grips with the initial ideas of ratio and proportion:

Grade three, the children just enter the classroom. The window is open. On the blackboard a big hand has been drawn. The teacher starts telling the following story.

"Once upon a time there was a giant. His name was Bert. He felt very lonely, because there were almost no other giants on earth. He couldn't find anybody to speak with. One day Bert wondered: would it be possible to talk with human beings? Sadly he bent his head: No, that would surely be impossible. People would be frightened to death. He was so big and ugly. But yet he couldn't forget the idea.

And so it happened that the giant Bert tried to make contact with people. He strongly hoped that human beings wouldn't be afraid. "...Look, here you can see what Bert showed the people... (teacher indicates the drawing on the board).

T: How do you know this is a giant's hand?

S: Of course it is. I never saw such a big hand before.

T: Do you think the giant is bigger than I am? (She asks this question to give a chance to the low achievers too.)

S: The giant's hand is much bigger than your hand. That's why the giant is bigger than you are. (The teacher emphasizes this statement by putting her hand onto the drawing.)

T: How big is the giant? Many children join the discussion giving their opinions: "As big as the classroom"... "a big as a tree". (But these answers the teacher didn't ask for.)

S: Look at my hand and look at the giant's hand. Compare both! (She puts her hand beside the big one, on the same line.)

Although the mathematical task for these third graders was obviously concerned with ratio and proportions, the setting of the task was a *non-mathematical* one. Of course this giant-context had been chosen for motivational reasons, and the pupils probably want to explore the situation, to comment on the teacher's questions which made them aware of the hidden problems." (*Wiskobas Bulletin* jrg 7 (1979))

But is this an appropriate context for the teacher to introduce the ideas of ratio and proportion? Is it meaningful or realistic for the children? Not many have met a giant, unless adults are 'giant' children. But does 'meaning' or 'realism' matter? What criteria should be used to determine the appropriate contexts for stimulating the development of mathematical ideas?

In the next example the pupils have to find rectangles of a given area K.

The pupils concerned already know something about the geometry of rectangles. They know the relation between dimensions and area.

There are at least two possible ways for them to tackle the problem:

- a numerical way – find 2 numbers a, b so that $a \times b = k$. The numerical field will depend on the knowledge of the pupils.
- Cutting and pasting – cutting n bands of the same width and pasting them together end to end so as to get a new rectangle. The numerical version of this procedure gives pairs $a' = a/n$ and $b' = b \times n$ so that $a' \times b' = k$.

We gave this problem to 9–12 year olds. These pupils were used to working with integers (addition, subtraction, multiplication and ordering, but no necessarily division). They used also some fractions in the sense of $1/n$ is a number so that $n \times 1/n = 1$ and p/n indicates the number $p \times 1/n$. We chose several values of k. The presentation below concerns $k = 24$ which is an integer with several divisors.

After having found all the pairs of integers (a, b) so that $a \times b = 24$, and having marked them on a grid, the pupils tried to improve their graphical representation. To do that, they needed not only to use fractions but also to find a process to create others. Indeed, they were convinced that between two marked points, there were a lot of other points corresponding to other rectangles with area 24. They put forward the following conjecture:

If a side is x times bigger, the other is x times smaller. So they justify (8,3) (24,1) (72,1/3)...

This statement formulated in this way is very general. However, some pupils question, at least implicitly, the extent of its validity. So, for example a pupil said:

"$(72\times1000, \frac{1/3}{1000})$ must give a rectangle with area 24."

But what does it mean?

An answer: "It is 1/3 divided by 1000, 3000 small pieces in the unit, it is 1/3000."

Another pupil said: "1/3000 is also 1/1000 divided by 3, it is $\frac{1/1000}{3}$."

So, we recognise here a "crucial experience (cf. N. Balacheff): to multiply and divide by 1000 are familiar operations. The generality of the statement allows such manipulation. However, the pupils find something which turns out to be unknown. After some reflection, they become convinced that the statement holds true. The crucial experience is not a mathematical proof, but the pupils use it to convince themselves of the extent of the validity of their statement."

A nice mathematical experience perhaps, but what makes us feel that?

How can we judge the worth of mathematical activity? Was it the richness of the problem which stimulated the mathematical activity? Was it the inventiveness of the pupils which made more of what might have been just an exercise in factoring 24? What did the teacher do to encourage the conjecture and its analysis? And what would a mathematics educator recommend as the *next* activity? Another area, like 36? Or the generalised problem, with area K? And why?

In the next illustration, Martin is a bright secondary school pupil, busy testing the teacher's (KH) definitions of the problem:

"K.H. "Give me two numbers (less than 6 and not 1, 2 or 4)"
Some lad "3 and 5"
K.H. "I am going to make up another number using multiples of 3 and 5, like: 17 = 4(3) + 1(5). Now you make me 12"
Louise "4(3) + 0(5)"
Martin "14(3) – 6(5)"
K.H. "No, you hou have made me tell you a rule, you can only add"
Martin "-6(5) +14(3)"
K.H. "No negatives at all"
Martin "Oh"
K.H. "Now make me 3"
Somebody "1(3) + 0(5)"
K.H. "Now 4"
There was a slight pause as young minds start to work.
Sarah "You cannot do it"
Martin "Can"
K.H. "How?"
Martin "1/2(3) + 1/2(5)"
K.H. "Positive integer multiples only"
Martin "You should have said." *(from Kent and Hedger, "Growing Tall", p. 167)*

Immediately one begins to wonder how the pupils understand this activity – is it a game? What are the rules? What is the right answer? Indeed, what does it mean that a pupil understands an activity? What do we mean when we talk about the understanding of rules? And what do we make of this teacher's role? What goals does he have? *He* seems to know because he evaluates the pupils' answers so quickly. But should the teacher tell the pupils what the goals are? Would they learn more if they shared his perspective? Perhaps an overview from the teacher would help: "Today we shall look at combinations of...?" Or would that defeat the activity? At another level, who is in control of the mathematical knowledge being negotiated in the classroom?

In a contrasting illustration, we now meet two secondary pupils, Linda and Jude, who were taking part in the LOGO Maths Project in London: In this project pupils working in pairs undertaking Logo activities as part of their mathematics curriculum. The computer is set up in the corner of the mathematics classroom and is used as and when appropriate. Linda and Jude had not come across Logo in their elementary school and their first introduction to Logo was when they joined the project class at age 11 years.

Throughout the whole of their first year of programming in Logo, Linda and Jude restricted their planned angle input to multiples of 45 which were less than or equal to 180. This strategy for angle input influenced the shapes they chose to produce i.e. squares, cubes, rectangular letters, and enabled them to cope easily with parallel lines or symmetrical slanting lines. At the beginning of their second year it was suggested that they draw some regular polygon shapes. They immediately drew a collection of different sized octagons! In order to provoke them to reflect upon their inputs to the turtle turns of RT and LT (Right Turn and Left Turn), it was suggested that they draw a hexagon. They agreed that a hexagon needed 6 sides and then, with no further hesitation, tried FD (forward) 40 RT 45 repeated 6 times. This produced an incomplete octagon much to their surprise. The computer output provoked the expected conflict in their minds and they adopted a trial and error approach in direct mode to 'discover' the turn they needed. Eventually they found that a turn of 60 'worked' and they produced the required regular hexagon.

At this point they were asked to draw a regular octagon – something they had successfully produced many times when using their RT 45 strategy. To use a 45 degree turn was not now their first inclination and Linda immediately made the following conjecture:

Linda: "Oh, that's excellent ... so what it is REPEAT 60 for 6 times ... so for an octagon it will probably be REPEAT 80 I'm feeling really brainy now..." So she tried her idea, that is 6 sides input 60; so 8 sides input 80 and she typed: REPEAT 8 (FD 50 RT 80)!

When asked why they had always used multiples of 45 for angle input before the episode described above Linda replied:

Linda: "Don't know ... cos we didn't want to explore any different angles I suppose."

Interestingly when using the 45 degree strategy Linda and Jude frequently 'tidied up' their turns, for example combining 3 turns of 45 to give 135. When Linda was asked (outside her Logo work) what would be the result of the calculation 45 times 3 she replied:

Linda: "Don't ask me ... I'm hopeless with numbers."

What was Linda's conception of input to RT and LT when using the 45 degree strategy? She recognised that RT and LT meant a turn which was to some extent ordered, but she made no synthesis with angle. Her work seemed to indicate a separation of an arithmetic frame for calculation, from a calculation undertaken implicitly within the Logo activity.

Further, what, one might ask, was the meaning to the girls of the activity of drawing an octagon, before the intervention of the teacher? What was the significance of that intervention? At another level, what is different about that teaching situation because of the presence of the computer? What is *its* role in the genesis of mathematical knowledge?

In our last example, we have an enthusiastic elementary school teacher faced with the fruits of his enthusiasm:

David Hansen was to have the 9-year-olds in his first year as a teacher. He decided to do a traffic project in mathematics. The class spent lots of time drawing road junctions and designing crossing for pedestrians. Some of the pupils positioned traffic lights as well, and at David's suggestion they constructed an algorithm for the timing of the light during the peak period and the rush hour. That made some of the pupils suggest that they could go out and collect some statistics on queuing at the nearby junction to see if the timing there could be improved.

Then Sue came up to him and asked him to sign a petition. "What's this for, Sue?"

Sue told him that the narrow road she lived in had been opened for through traffic. Even the lorries used it. The families living along the road wanted it to be closed to through traffic.

"We cannot play there any more. And when mother pushes the pram she has to hide in the hedges when the big cars are passing by. Father and Mr Brown have made a drawing of the bend. Father claims that it is impossible for the lorries to pass the bend without getting into the hedges. Mr Brown says it is possible if the drivers use the whole width of the road. You couldn't imagine how they quarrel. They make models and push toy cars all evening. Would you sign Mr Hansen?"

A clearly relevant mathematical activity for Sue but a dilemma for the teacher. How far should the activity venture from the 'safety' of the classroom? Should he use the opportunity presented by Sue? What is next on *his* mathematical agenda, and how should he deal with her request if he decides not to continue with traffic projects? On the other hand it seems easy to guess what is on her mathematical agenda, but can we really know? Does it matter if our conjecture is wrong? And what of the other pupils' agendas?

DIFFERENT QUESTIONS, DIFFERENT PERSPECTIVES, ONE COLLABORATIVE RESEARCH PROJECT

Five classrooms, five teachers, five groups of pupils and five different contexts for learning mathematics, for developing mathematical knowledge. And each situation raises many questions for mathematics educators. Too many sometimes, so how does one decide on what are the most important and most significant questions to tackle? That, in essence, has been the BACOMET group's task in this project. Classroom situations have provided our contextual focus, and our search was for the similarities at issue – in a sense anyone can see differences between the five situations just presented – the ages of the pupils, the mathematical topics, the sizes of the social group involved, the stages of development of the knowledge, the gender of the players. But the real didactical task must focus on the similarities – what problems do classroom situations in learning mathematics have in common? What issues are always present irrespective of milieu, content, players? And what, of these, are the most significant for developing mathematics educators' understanding in order that mathematics education itself can develop fruitfully?

The BACOMET group, with its members coming from different countries and raised in different research traditions, tries to maximise and exploit the tensions between the interpretations arising from our different backgrounds. We believe that this process of confrontation of different research perspectives is crucial for the identification and analysis of the really significant aspects of the process of the growth of mathematical knowledge in the classroom.

The most obvious aspect of similarity concerns the nature of the mathematical knowledge with which we are dealing – what is specific about it and what marks it out *in the classroom* as a particular kind of kowledge with its necessary activities, criteria and structure. The 'external' features, as van Dormolen (Ch. 4) calls them, are revealed as problematic in the first and last examples above – not in the global sense of assessing the total contribution of mathematics to society, but in the more localised, classroom-based sense of evaluating and judging the relevance and 'worth' of mathematical activities in the context of individual children's lives. We must recognise as well that teachers have a responsibility both to the individual children they teach and also to the society in which they all live, and that tension must be reconciled somehow.

The 'internal' features of mathematical knowledge are revealed as problematic by the second, third, and fourth examples above. The connectedness, the relatedness, the flow, the sequencing, are all sources of problem for both teachers and pupils. How can something as complex, and as fragile as mathematical knowledge, as Otte and Brousseau analyse it (Ch. 1), be structured so as to be accessible and knowable by children? Does the development of metaphoric knowing, as discussed by van Dormolen, hold strong promise for mathematics education? How far can Douady's (Ch. 5) elements of tool/object/setting/window go in providing a conceptual framework for teachers and teacher educators? The choice and nature of the mathematical activities in the classroom are clearly important issue areas in relation both to the teacher's range of mathematical knowledge and to the pupils' mathematical development.

Another aspect of similarity in the examples above reveals the pupils trying to make mathematical meanings for themselves. Martin, and Linda, are most clearly trying to do that either alone with the teacher, as Martin in effect is, or as a partner in a pair, as Linda is. Collaborative meaning-making is nicely exposed in the second example concerning rectangles, where we can see the fascinating 'meaninglessness' of mathematical idea-stretching weaving its magic spell! So how can didactical analyses help mathematics educators in this problematic area? Does the analysis of Dörfler (Ch. 3), with his emphasis on the key process of generalizing, help us not just to interpret some of this meaning-making but also to advise teachers on how to help children with this proces? How far should we heed Werner's call (Ch. 6) to observe, observe and again observe the children, in order to see the rich meanings they actually make for themselves? Does Hoyles' (Ch. 7) concept of the 'microworld' offer us a promising new perspective on the pupils' construction of mathematical reality?

At the heart of the teaching/learning process is of course the teacher, not alone with each pupil but in a social milieu with its own structures, tensions and history. That milieu is ever-present and all important, and we can see the dilemmas and challenges for the teacher which are illustrated by the 'rectangle' example, by David Hansen's situation, and by anticipating the intriguing future of the class who will come to discover much more about the life and nature of Bert the giant. Could these teachers benefit from understanding more about the pitfalls presented by the double-bind phenomenon as analysed by Mellin-Olsen (Ch. 2)? Can a greater awareness of the mathematical values which are developing in any mathematics classroom, according to Bishop (Ch. 9), help teachers to see what lies behind some of the social tensions they have to handle? Could more research like Balacheff's (Ch. 8) reveal to mathematics educators and teachers the strengths and weaknesses of different kinds of social interaction patterns?

Questions like these the readers must answer for themselves, for there are no neat solutions in this book. Nor do any of the chapters focus on only one component of the process and ignore the others. Whether the discussion is about metaphors, social interaction, computer microworlds, conceptual complexity or paradoxes, each of the BACOMET contributors is fully aware of the complex interrelationships in the classroom between the mathematical knowledge, pupils' meaning-making and teachers' knowledge-shaping. This means that there can be no instrumental or pseudo-logical sequence of chapters. The chapters do relate however, and one set of relationships has already been mentioned. Others however exist and another sequence of chapters has been chosen, with an Entr'acte between successive chapters to assist the reader.

The first problem for the BACOMET group was to find the appropriate unifying theme for their work and for their collaborative and personal writings. 'Knowledge' was the first focus, then the 'Didactical situation' presented another thrust. Finally the 'Genesis of mathematical knowledge' brought the two earlier foci together. The last problem for the group was how to create this book, so that the mathematics education community might benefit from the work of the group. Like the analyses and reflections with which the individual members have challenged each other, the book does have a beginning – this Introduction – but it has no end. This reflects the work of the BACOMET group – the book in no sense represents the end of the task.

ENTR'ACTE 1

This first chapter, by Otte and Brousseau, introduces us to the fragile quality of mathematical knowledge. In a world where mathematical 'facts' are rarely questioned and where mathematics in most people's minds is associated with the 'hard' sciences of physics, chemistry and biology, this notion may be rather a surprising one, particularly when meeting it in the first chapter. That is precisely the point of this first chapter, to challenge the reader not to accept the seemingly acceptable language of educational mathematics, not to remain at the superficial 'conversational' level of describing and labelling mathematical knowledge. For those who work in education, particularly, it is necessary to question the everyday understandings in order to reveal the dilemmas and paradoxes in the processes of mathematics education. Paradoxes figure fundamentally in this chapter, not for their own sake, but because of their particular metaphoric significance. The teaching and learning of mathematics is a social act, first and foremost, and at its heart is what the authors call the Topaze Effect – the impossible task of the teacher to try to teach what the learner cannot learn. But the phenomenon of the didactical contract demonstrates that the learner is also in an impossible situation, trying to learn what the teacher cannot teach. So how can didactical analysis help us to come to terms with these impossible statements? How can mathematics education progress in the face of these paradoxes? How can mathematics educators respect the fragility of mathematical knowledge in order to help learners gain access to it?

GUY BROUSSEAU AND MICHAEL OTTE

THE FRAGILITY OF KNOWLEDGE

INTRODUCTION

The social practice of knowledge is conceived of either as the administrative problem of product distribution or as a question of individual access on the consumers' side. The notions and mechanisms through which one handles these questions are of proverbial debility and yet thoroughly accepted. It seems that never before had knowledge been so readily available, but it is at the same time extremely fragile. The fragility of knowledge is accompanied by increasing difficulties of handling knowledge and of establishing relations with the practice of knowledge.

We don't intend to repeat here all the well-known difficulties related to mathematical knowledge. Rather we should like to propose quite a different interpretation of these. Our dependence on (mathematical) experts, for instance has become a much touted problem nowadays. The abstractness of mathematical and scientific knowledge is considered responsible for the fact that the educational system fails to produce at least competent questioners, cross-examiners and adjudicators of our experts. In reality these considerations are a long way from the point as we have already completely instrumentalized our experts along with their mystification. Every political party or economic pressure group has its "own" experts. We have in fact lost all sensitivity for the fragility of knowledge and the politics of truth in our society.

Didactics naturally is led to investigate such questions, but generally in a more technical or management-like manner. Didactics, however, must always operate on different levels which cannot be fully integrated. The practical approach is very often blind to the phenomena which make up the central concern of this article: the didactical contract is paradoxical; both intuitive and formal control of the signification process have their limits; creation and communication of knowledge appear as inseparable and yet incompatible with each other; every creation of meaning is nothing but the starting point of a new effort towards the same goal. ...

All these phenomena can be observed at the level of individual teacher-student interaction, but they appear at once to be mere singular expressions of more general realities. It is important for the didactician and for the teacher to see the appropriate frame because the phenomena are complex and elusive. If the teacher for instance concentrates on himself or on an in-

Alan J. Bishop et al. (eds.), Mathematical Knowledge: Its Growth Through Teaching, 13–36.

dividual pupil, any meaning of the knowledge in question evaporates. Every thought demands an appropriate "space" of social practice to be meaningful. Hence we know much which we cannot really use and our decisions, on the other hand, are not considered to be related to our knowledge.

Our dealings with knowledge lead to abrupt oscillations between a factual concept of scientific knowledge as direct truth on the one hand and a concept of knowledge as a mere vehicle for arbitrarily chosen goals on the other. The project of the encyclopedists – who attempted both to integrate scientific knowledge into our culture as well as to make a scientific culture, is randomly replaced by activities, which while losing all interest in the referential aspects consider knowledge to be but a question of control, and see meaning as but a matter of usage. The didactical programs have followed these fluctuations very closely, thereby signifying firstly that questions of epistemology conceived as the politics of truth lie at the heart of the matter, and secondly that these questions are concerned with the paradoxes of a simultaneous description and construction of individual and social reality.

Two fundamental insights can be gained by scrutinizing the development of mathematical education during the 20th century (in fact they may with little effort be extracted from the existing literature, for instance from books like W. Wooton's "SMSG, the making of a Curriculum", 1965 or from similar ones).

First the fact that the truth, and the social function, of scientific knowledge are on the one hand inseparably linked, while on the other hand they cannot be reduced to one another. It becomes particularly clear that the social functions of science, such as education or technological application cannot simply be derived from the truth or objectivity of the knowledge and that there is no singular mechanism (like the idea of "structure of the discipline" or the "fundamental concept" of set etc.) that might help to "overcome" this fact.

On the other side, and here comes the second conclusion, lies the utilitarian viewpoint, that the influence of the pragmatic philosophy, which has reigned in our mathematics classrooms since the turn of the century has failed to produce satisfactory results through ever more reducing the goals of mathematical education. More and more emphasis was placed on "useful" mathematical procedures and less and less time was devoted to theoretical and meta-theoretical considerations. Despite this the level of proficiency even in this reduced area dropped ever more and remedial mathematics courses became an everyday part of schooling.

The same situation can be observed with respect to technological application. On the one side there is at the moment in our society a trend which tries to limit scientific research to the economically useful domains and to relegate questions of the cultural role of science in its broader concerns to the background. On the other hand however we see that research policy is not able to find acceptable guidelines and orientations on such a narrow basis nor to decide on alternatives in research in such important fields as the future energy supply or the space program.

We have to realize that the character of to-day's scientific knowledge and the problems of its social practice demand very comprehensive orientations, even if we are concerned with very specific and detailed questions. The singular and the universal in cognition are inseparably linked and at the same time again they cannot be reduced to one another. On the surface, education has for years been occupied with the "fragility" of the pupil. It now seems necessary to become aware of the fragility of knowledge, for teaching and learning is, after all, concerned with knowledge. This might even help us to understand better as well as to practice the social relations of those participating in the politics of truth, especially as the fragility of knowledge is just one expression of the contradictions of "modern life".

We shall in the paragraphs which follow just provide some seemingly disjoint arguments which hopefully will nevertheless contribute to building up an intuition of the theme.

2. A DIALOGUE

To explain the title, we present a dialogue of two persons (two teachers perhaps) which is, let us say, stimulated by watching a classroom situation. The situation itself is taken from Marcel Pagnol's famous play "Topaze".

The first scene in this play illustrates one of the basic processes of schooling: Topaze is setting a dictation for a weak pupil. Unable to accept errors that are too gross and too numerous, and not being able to give the required spelling more directly, he "suggests" the right answer by hiding it under increasingly transparent didactic encoding: "...des moutons étaient réunis dans un parc ..." problems of spelling and grammar beset the pupil "...des moutonsses etai-hunt...". The problem is completely changed! Confronted with repeated failures, Topaze begs for the correct endings and negotiates on a falling market the conditions in which the pupil will eventually put this "s". We imagine that he could continue by requiring the chanting of the rule, and then require it to be copied out a certain number of times. The complete collapse of the act of teaching is represented by a simple order: put an "s" on "moutons": the teacher has in the end taken over what was the essential part of work".

What is wrong with it?

- What are the assumptions? It seems that the pupil's inability to detect the solution forces the teacher to *give* it to him.
- If it is, however, *possible* to hand over the knowledge to the pupil, why should the teacher not do so? What is the question posed by the Topaze Effect?
- The Topaze Effect assumes that it is impossible, for instance, to hand

over the meaning of a theoretical concept to the pupil like a concrete object. Concepts cannot be taught directly, because the pupils have to reconstruct the meaning for themselves.

- But if concepts (and the like) are a strictly personal thing, why do we have concepts at all? What is their use and why should they be taught?
- Would this view not eradicate all distinctions between theoretical knowledge and everyday experience? Would it not even destroy the purpose of general education, because practical skills and everyday experience are better learned by doing, and are more easily acquired on the job? Has not general compulsory education, as institutionalized in our schools, always been dependent on the theoretisation of knowledge despite the fact that this theoretisation permanently causes pedagogical troubles, as we have just seen?
- I am not at all impressed by your lengthy plea. In fact I am going to strengthen my claim that theoretical terms are dispensable. If they do not serve their purpose, they are dispensable, okay?
- Okay, but ...
- If they serve their purpose, they establish relationships among facts. If I know these facts, the same relationships can be established without theoretical terms. Thus, these terms are dispensable.
- The essential distinction should be made between object and process, between the given situation and time. Both personal knowledge and social knowledge are processes which have to be put on the stage like a theater play. With all learning, even with the simplest rote learning, there inevitably goes along some meta-learning, some "learning to learn". Theoretical terms serve this necessary meta-learning.
- Oh, I know this kind of reference to the "meta"-level quite well. Usually it is employed to argue that a machine, for instance a computer, cannot think because it must be instructed in a direct way and cannot be provided with the necessary meta-rules for applying the rules. But the lowest level rules (or in case of humans also the highest level rules) are built into the hardware and function all by themselves according to the laws of physics (or of social history). We are not getting further in our discussion. We started with some pedagogical problems related to theoretical terms. We went on to the epistemological and ended up with the question of the difference between humans and machines and with the question of whether machines can think. It seems...
- Again you did not listen carefully. The meta-level as such is not important but the *simultaneous* presence of different roles is. To understand some piece of knowledge means to simultaneously experience the knowledge *and* some application of it. In the main, this simultaneous experience of structure and time, of object and process is provided by social contact and human cooperation.
- Is not the teacher pupil relationship an example of this and ...

– Yes it is. This contact however cannot be reduced to explicit communication, as we obviously know more than we can tell. Teachers influence their pupils not only by what they explicitly do or say, but also by what they are as persons and by the kind of intellectual life they themselves live, by the kind of commitments they make. Is not the transmission of knowledge from one generation to the other anyway predominantly tacit?
– Stop these exagerations, before you plunge completely into a disgusting traditionalism. Scientific knowledge cannot be reduced to shared beliefs or common intuitions. Is not mathematics in school the most important means to foster independent and critical thinking? Has it not always been an instrument for gaining new and unexpected knowledge as well? By intuition, something is only given to us, not *apprehended*. This intuitive knowledge, in fact, would be a contradiction in terms. Even a person's most intimate experience is a conscious attitude – and is not mere conditioning or indoctrination – only in so far as his dealings with it involve explicit knowledge.
– This might all be correct, but may I ask you a question? Do children in reality learn to think critically and independently through explicit mathematics instruction in school? For the most part they do not, because they are not motivated. They have no problem in which to invest their mathematical activity and such a problem cannot be given to them just by explicit communication and formal contact. No new knowledge can emerge in this way either, because explicit communication has to reduce the new to something that is already part of the mental stock of the receivers. How could they learn something radically new in this manner? And as to my alleged traditionalism: have not the great inquirers always had the habit of bringing their singular conceptions into line with the general conception or intuitive idea of an entire province of phenomena; a province which is so large that it could never be the subject of a scientific theory in the proper sense?
– It seems to me now that a theoretical term would at the same time be the starting point and the goal of cognitive activity. It seems that the intuitive knowledge can be functional for attending to the related explicit definition, and the latter serves to transform vague intuitions into personal experience ... This is circular ...
– Indeed! Not to judge it to be a vicious circle, we have to keep in mind that learning consists of independent activity, direct instruction and a general process by which children grow into the intellectual life of those around them. Especially the teacher-student interaction cannot be reduced to explicit communication and direct instruction – which in fact may be delegated to "teacher-proof" curricula or computer programs – but has to rely on extensive personal contact.
– I am not completely satisfied. It seems to me that you eliminate the "problem of knowledge" by relegating it to the psychological and the

social levels.

– That is not really true. I have said before that school as a social institution is based on the theoretisation of knowledge, and that the teacher-student relationship is unthinkable without considering questions of knowing and learning. It is in fact based on a kind of didactical contract.
– This explanation does not help me very much. Nevertheless I have at last now understood that the term "knowledge" is as elusive a concept as the term "I".
– Exactly, and for the very same reason: because of its fragility.

3. THE DIDACTICAL CONTRACT

The contract: the teacher is obliged to teach, and the pupil is obliged to learn. Upon considering this definition of the contract, some tend to say: there is nothing meaningful about it. It's just a definition of the terms of "teacher" and "pupil" respectively, which we know already. Others understand the contract literally and absolutely and hence raise objections, saying it demands too much, indeed, taken absolutely, the impossible. Taken together, these reactions show the very essential nature of the above characterization of the teaching-learning situation. They must be taken seriously, as they describe a kind of ideal reality, but they must not be taken literally in an empirical sense.

For the persons involved, the contract is mainly implicit and it cannot be negotiated. Hence teacher and pupil are all the time busy inventing ever new forms of behaviour and interaction, which they hope to be in accordance with the contract, being either interpretations of it, or tolerable evasions. Neither of the two parties is able to completely control the contract, nor may it be ignored.

The contract must be honored at all cost, for otherwise there will be no education. Yet to be obeyed, the contract must be broken, because knowledge cannot be transmitted ready-made and hence nobody – neither the teacher nor the pupil – can be really in command.

In mere training, there is no such contract. Judah Schwartz makes a similar "distinction between education and training in order to underline the fact that the nature of the social contract between teacher and learner is different in the two cases. In training, there need not be an explicit or implicit obligation for the teacher to teach or for the learner to learn the underlying intellectual structure of what is being taught and learned". In Schwartz' (1983) view, "education in a democratic society must be based on a different set of assumptions about teachers and learners. The price of a free people not knowing why it believes what it believes can, in the end, be the loss of that freedom".

What does this mean however? Is a driving instructor not obliged to teach

... etc.? To teach or to learn, one needs methods. "Method" labels the seemingly unproblematic meeting ground for teacher and pupil. This very often leads to "overmethodization" which is caused both by the teacher's desire to channel the child's activities ever more narrowly, and the child's demanding definite rules and an ever narrower specification of the things to be learned.

Only in the case of very few forms of knowledge, for instance when I want to learn how to drive a car, are the methods of learning and teaching identical with immediate application. Only in such cases can the process of application be specified by the method of instruction. Theoretical knowledge is not of such a kind.

We are now able to answer the above question about explicitness. We have stressed the systematic, and by no means accidental, role of intuitive or tacit knowledge. We have also explained that the relevance of intuitive or tacit knowledge must be seen related to the part a person consciously appropriates for himself; an achievement which explicitness and an independent critical attitude. To give an illustration: a keen soccer player will probably be aware that some prominent goalkeeper did not become famous in soccer because of his mathematical ability which told him where a goalkeeper should stand to narrow the angle. If, however, that person wanted to be as good a goalkeeper or to do even better than his shining example, he would necessarily have to rely on making explicit and evaluating critically what the other attained intuitively and through experience.

Let us repeat that this does by no means imply that tacit knowledge can completely be made explicit, nor that intuitive beliefs do not play an essential role in personal development. Tacit knowledge is in fact indispensable because there is no possibility of each succeeding generation, let alone each individual, critically putting all the teachings in which he is brought up to the test.

What then about J. Schwartz' claims with regard to the essentials of education in a democratic society? Has scientific rationalism not been a major influence toward intellectual, moral and social progress ever since the French Revolution and until our own days? That is true. Equally true, however, is the observation that a scientism which confused scientific knowledge with the reality to which it is related, has resulted in harmful social practice and false ideologies. Theories neither present us with a set of bare facts, giving humans an unrestricted choice of the rest, nor do they provide a justification for practices which try to lock the human subject into ideological traditionalism or utopian moral perfectionism.

Scientific theories consist of factual knowledge and at the same time their meaning and their relevance is based on the yet unintended and unexpected applications. They are teleologically due to the objective and technical purposes served by them and are yet autotelic, thereby serving an open self-development of the human subject. It has been repeatedly shown in philosophical argument that subjectivism (which ignores the fact that the concept of the human subject cannot be separated from the project of its evolution) and empirical or logical objectivism (which forgets that the foundations of knowledge cannot be sought outside its growth) are but two sides of one coin (see, for instance, Popper's argument against Hume in his "Objective Knowledge", 1974).

At first glance, it may appear that awareness and action stand in antithetical opposition in the sense that a person who understands everything will be unable to take a decision. Theoretical understanding of reality is in fact based on such an opposition and conversely defines its unity as an evolutionary process, that is as the development of the knowing subject himself.

LaPlace's "infinite intelligence", while aware of every particle of the whole universe, nevertheless does not possess any reasonable knowledge, because this infinite intelligence itself is no longer a part of that universe. As a knowing subject, it has no place in that universe and its knowledge of it has therefore no purpose. On the other hand, somebody completely contextbound like one of the material particles would also be uttelry unable to gain insight, because insight presupposes the subject's ability to see herself in relation to an "outside" reality. Without such an insight knowledge would again lack all purpose.

It is therefore the growth and development of the knowing subject as an integral part of some contexts which provides knowledge with sensible foundations.

– If knowledge is but a set of rules, methods or algorithms to be directly applied, the subject (teacher or pupil) is not involved, nor will he/she be transformed (developed) by the cognitive process. Rather, she will only be conditioned by the context of application.
– If knowledge is but emotion, feeling or awareness, she may become involved in the sense of becoming encapsulated and locked.

The complementarity of method (form, style etc.) and application (content ... etc.) characterizes theoretical knowledge. Together, the complementary processes of formalisation (giving form) and of application (giving content) constitute the personal knowledge of the cognitive subject and thereby his individuality within the context of social history. The fragility of knowledge and the difficulties of the didactical contract are essentially linked to each other, and result from the openness of history.

The following remarks and quotations from Wittgenstein (1967) and Bateson (1973) intend to assist didactical reflection on these matters.

"It might be asked: What is there about the proof of Dedekind's theorem that a child 10 years old could not understand? – For this proof is far simpler than all the calculations which the child has to master? – And if now someone were to say: it can't understand the deeper content of the proposition – then I ask: how does this proposition come to have a deep content?" (Wittgenstein, 1967, p. 148e).

The answer again is: by means of applications, by means of the *unexpected* applications. Applications which are therefore chosen and decided on by the learners themselves.

This results in the "paradox of the didactical contract" between teacher and learner. If both the problem and the information about its solution are communicated by the teacher this deprives the pupil of the conditions necessary for learning and understanding. The pupil will only be able to reproduce the method of handling and solving the problem communicated to her. Again, as in Wittgenstein's example, we are led to observe that mathematics is not just a method. We do know, on the other hand, that isolated problem situations do not of themselves produce means for their analysis or solution. And, in addition to that: what would be the relevance of a specific application of knowledge or of the solving of a particular problem?

It is obviously impossible to understand the crucial function problems have for the development of cognitive activity without considering "problems as variables" and, accordingly, searching for the mechanism regulating this variability. Finally, it is within the logic of the idea of "problems as variables" to see the crucial point in the connection of objective and social relations. Our interactions with objects are generally also interactions with others, and a problem is neither a set of statements nor a purely objective, isolated entity, but it is an objective as well as a social relation described by a set of statements. To put it differently: education and development (Bildung) is the central notion of didactics and pedagogy, not "learning", "problem-solving" or even "creativity".

Searching for the "regulatory mechanism" of some variability amounts to looking for a meta-perspective. From such a perspective, G. Bateson (1973) has presented a complex hierarchy of processes of learning where a "profound reorganization of character" takes place on the highest level of "Learning III". Stated differently: the problem to be ultimately "solved" or the "application" of knowledge intended in the end is the human subject himself. This self-application or auto-development is, however, of a contradictory or even paradoxical nature, as we shall see. The contradiction lies in the fact that real human subjectivity or selfhood has both an individual and a social character. It is in particular both an individual mind and a "society of minds" simultaneously.

Let us recapitulate our exposition of the "paradox of the didactical contract" by means of Bateson's afore-mentioned hierarchy of levels of learning and meta-learning. This hierarchy is based on a formal argument which realizes that a meta-perspective consists of transforming some fixed "constants" (contexts etc.) into variables. Generalization amounts to the introduction of variables. Bateson summarized the hierarchy as follows: "Learning I is change in specificity of response by correction of errors of choice within a set of alternatives. Learning II is change in the process of Learning I, e.g., a corrective change in the set of alternatives from which choice is made, or it is a change in how the sequence of experience is punctuated. Learning III is change in the process of Learning II, e.g., a corrective change in the system of sets of alternatives from which choice is made." (p. 264)

A classical example of Learning I is provided by the behavior of a "Pavlovian dog". We might also say that reproducing the proof of Dedekind's theorem as a result of rote learning is an example of Learning I. We might, however, realize that, especially in mathematics rote learning does not exist without the occurrence of "learning to rote learn", which obviously represents an example of Learning II. "What the data show is that for any given subject there is an improvement in rote learning with successive sessions, asymptotically approaching a degree of skill, which varied from subject to subject. ... This improvement cannot have been due to Learning I. ... It is also worth noting that educators have strong opinions about the value (positive or negative) of training in rote learning. Progressive educators insist on training in insight, while the more conservative insist on rote and drilled recall" (Bateson, pp. 265/266).

It is rather unclear how Learning III may be described positively and it is difficult to give examples as they occur very seldom. But to apply the same sort of logic to the relation between Learning II and Learning III as was used to describe the relation between Learning I and Learning II, already leads to the essential point. "Learning II is an economy of the thought processes (or neural pathways) which are used for problem-solving or Learning I. The premises of what is commonly called 'character' the definitions of the 'self' – save the individual from having to examine the abstract, philosophical, aesthetic and ethical aspects of many sequences of life. But Learning III will throw these unexamined premises open to question and change." (Bateson, pp. 247/275)

4. INTUITIVE AND FORMAL CONTROL OF MEANING

4.1. In December 1967, G. Papy gave a lecture at the Düsseldorf Academy of Science in which he expressed the hope that with the advent of New Mathematics intuition and rigor will be in harmony for the first time since the days of Euclid and the new learning will be active, motivated, and scientific i.e. in strict accordance with the rigor of modern mathematics. The basis of this harmony of intuition and logic should be described as "formal". The "formal" is seen here in contrast to the material – a piece of music is formally the same despite the variety of its material performances – as well as in contrast to the context-bound or to the merely algorithmic or mechanical. There may be people who were drawn to mathematical logic because they had an obsession with making everything entirely secure even for a mechanical intelligence. It is however a familiar misconception, to believe that mathematics or mathematical logic is to be engaged primarily in such mechanical thinking.

The idea of "formal"intended here is linked with a conception of meaning as related to structure and not to reference nor to mere use. "Colourless green ideas sleep furiously". Even such a sentence (it was given by Chomsky to illustrate the relatively independent existence of form and meaning and has been quoted since 'everywhere'), even this sentence 'makes sense' when interpreted metaphorically. Thus, metaphoric understanding is formal understanding, as is all theoretical understanding. The formation of a science results from singling out a set of concepts and laws which formally hold together, as well as a method of explanation, or logic, which renders such a science autonomous, that is independent of all material relationships with any other domains of experience, despite the richness and usefulness of such relationships as in applications. (This idea of concept formation in science has most clearly been stated by G. Cantor, 1932, p. 201.)

The intuitive aspects of scientific thinking, as opposed to the aspects related to its method, are those who refer to meaning in the sense of a feeling that the theoretical objects about which one speaks really exist. In this sense, R. Thom said at the Exeter Congress in 1972: "The real problem which confronts mathematics teaching is not that of rigour, but the problem of the development of 'meaning', of the 'existence' of mathematical objects" (Howson, 1973, p. 202).

One should, however, not forget that this existence is theoretical or mental existence, not concrete existence, just as the map is not the territory. It is based on formal meaning in the sense described above. With reference to outside reality, one might instead say it is based on metaphorical meaning. In fact, Jacobson and Halle (1971), in their analysis of speech disturbances, identified one essential type of such disorder as the inability to produce pure predication, which they showed to be a disorder of the metaphoric way of developing discourse.

This type of aphasia is, for instance, characterized by the inability "to produce a purest form of equational predication. The tautology $a = a$: or 'no' is 'no'" (p. 18). A patient with this type of aphasia doesn't utter "the word 'knife' alone, but according to its use and surroundings alternately calls the 'knife' 'pencil-sharpener', 'apple-parer', 'bread-knife', 'knife-and-fork', so that the word 'knife' was changed from a free form, capable of occurring alone, into a bound form" (p. 79).

"The aphasic defect in the 'capacity of naming' is properly a loss of metalanguage. As a matter of fact the examples of equational predication sought in vain from the patient's cited above are meta-linguistic propositions referring to the English language" (p. 82). It is quite natural that patients of this type could not be brought to understand the metaphoric character of words. Metaphorical meaning does not exist if meaning is reduced to use in actual situations.

"The theory of language since the early Middle Ages has repeatedly asserted that the word out of context has no meaning. The validity of this statement is however, confined to aphasia or more exactly to one type of aphasia" (p. 79), namely to the absence of the metaphoric.

Textbooks as well as popular science reports lavishly use metaphors, but do not provide any real insight because of the arbitrary and inconsistent way they are using them. They ignore the fact that real knowledge has an existence sui generis based on a structure of relationships. Or their authors do not believe in the truth of their own metaphorical explanations because they confuse knowledge with absolute certainty. To identify the quest for knowledge with the quest for certainty has been called the weakest part of the common sense notion of knowledge by Popper. In fact, metaphors in textbooks are then no longer understood as metaphors, but are taken literally. The arbitrary looseness all of a sudden turns into the opposite and these texts obsessively aim at arousing a precise response on the part of the reader. But they cannot be understood in this way. They ask for ever new explanations and the presentations become over-methodized.

This is a real paradox: a "formal" structure is necessary for an open text and the productive use of metaphors. Within a speech act theory, an understanding of the real purpose of the metaphorical can never be developed! If the metaphorical is absent, meaning is identified with "use", i.e. is completely context bound. An open text, based on the semantics of metaphor, presupposes a specific competence on the side of the reader and at the same time develops this competence.

Metaphorical understanding is based on intentional ontology: intentionality refers to consciousness in its active grasp of objectivity. A metaphor consists of an object *and* a specific way of representing it as part of a certain structure. Its not the object, let's say a "set" but this "set" taken as the content of some activity and consciousness. In this way a metaphor presents the well-known as something new and fundamentally important. Metaphor seems to lead to the essence of the things themselves. Metaphor is very often confounded with analogy. Analogy is based on structural similarity. Two things "are analogous if they agree in clearly definable relations of their respective parts" (Polya). Metaphor cannot be reduced to structural similarity alone, but is, as was stated already, also dependent on the manner in which something is given to the mind. Metaphors are mentally constituted by their function of coordinating the category of objectivity and the category of evidence. The functions of metaphorical language, especially in mathematics are like those associated with perception. This fact links metaphor with the problem of a holistic ordering of a new intellectual field and accounts for the possibilities offered by metaphorical expressions of epistemic access to the world in a way not replaceable at the time by any literal expressions. So metaphor is closely linked with theoreticity of our world view as such.

A different perspective on metaphor stresses its role in social discourse and sees it as an important mechanism in the cultural fixing of new meanings over time. Taken together these views promote the notion of metaphor as a link between the problems of foundation and of growth of knowledge. Metaphor in relating the question of foundation of knowledge to the other aspect of the dynamics and evolution of knowledge at the same time shows that these relationships cannot possibly be established without accepting that knowledge and the objectivity of knowledge are social or cultural artifacts.

One might now pose the following question: did the New Mathematics reform movement fail because of the difficulties of acquiring a "theoretical attitude" and accordingly see that the fundamental mathematical concepts like the notion of "set" were to be understood metaphorically or did it fail because of the inappropriate choice that was made with respect to its fundamental metaphors?

Papy's hopes of bringing intuition and logic, evidence and objectivity into harmony was based on his belief that there existed a notion equally current in everyday thinking as it is fundamental in pure mathematics, namely the notion of "set". Did he fail because he (and the whole movement of the New Mathematics reform as well) had forgotten that this notion of set was to be understood metaphorically in order to relate the procedural everday understanding and the "formal", that means phenomenological understanding? Or was the notion of "set" the wrongly chosen metaphor? (Was it, for instance, much to little specified?)

If to-day people who had been active in the New Mathematics movement are asked, what according to their opinion has been the reason for its failure, they usually point to the first type of argument. They might say for instance: "We have learned that only a minority among our pupils is able to grasp theoretical mathematics." As general education is always dependent on the theoretisation of knowledge, this answer implies that there can be no really general mathematics education. The New Mathematics reform movement had however aspired to exactly that: mathematics for all!

To us it seems that the answer must be sought in the other direction. It seems in particular to be a wrong headed idea to look for just one fundamental metaphor as universal foundation. Metaphors must be knowledge based and must be able to be employed in a number of variations by different people. Our hypothesis is supported in particular by the fact that in New Mathematics the means of presentation were turned into new objects of instruction. As the school textbooks and curricula obsessively aimed at arousing a "precise understanding" in the sense of a definite response on the part of the pupil, they became a fact open to any possible interpretation. Only the reliance on a "formal" understanding in the sense outlined above, i.e. an understanding that would actively decipher the meaning of any textual element from the structural contexts into which it might be embedded, could have avoided this detrimental vagueness.

4.2. Bourbaki's "fascicule de résultats" of set theory begins with a statement which specifies its role for the student: in it, the latter will find "all the definitions and results ... used in this volume" and will be able "to confine himself to attributing to them (to the terms introduced without definition) their usual meaning ... which makes the propositions advanced almost self-evident". Hence, "the results of set theory", presented together with their formalization in the canon, are most certainly an object of teaching. The means suggested to the student for mastering the use of this theory is the "usual meaning", that is, on the one hand, a general and somehow ontological metaphor, and on the other hand a "literal use of the definitions of the technical terms" (Godement, 1963, p. 19). In fact, much care is devoted, in genuine textbooks, to presenting a large number of exercises, "the lack of which would reduce the utility of a book intended for beginners to almost nothing" (*ibid*, p. 17). Mastery of the conjectures in this fundamental chapter is thus mainly semantic. The proofs of the most difficult theorems and the difficulties of logic will be examined in another volume (book one) and from a "formalist" point of view, the view of the "fascicule de resultats" of being declared "naive". Hence, the students are expected to check the validity of what they put forward in this field through their possession of a "representation" of the things about which they talk.

Sets are a sort of substitute, a universal metaphor for mathematical structures, but a large number of exercises are in fact necessary to distinguish the "rightful" uses of that "language" from those which are beset with gross and unsuspected contradictions.

What we see upon comparing the naive use of the idea of set with set theory as a formal discipline, and upon noticing that in Bourbaki both occur with somewhat indistinct borderlines, is that intuitive and referential meaning and technical sense in modern mathematical theory cannot be brought into positions which are clearly separated from another and static. Mathematics always uses metaphors and tries as well to regulate and control the intuitive meanings. To say it in other words: foundation and growth of knowledge become more and more intertwined.

Pure mathematics has used the notion of "set" as a universal ontological metaphor in an attempt to keep its unity. The logical paradoxes have clearly shown that methods which were everything but intuitively clear and simple had to be used to make this ontology work (besides, these attempts were not very successful). For the mathematical researcher, this ontological metaphor has always been embedded within the structure of a mathematical style expressing mathematical practice as theoretical activity.

Symbolic models or representations do not liberate us from the necessity of having operational systems and intuitively accessible ontological foundations simultaneously (which should both evolve in a parallel fashion), or, in terms of school mathematics, of having formulae and applications. The representations themselves are vehicles of this double movement. In any case, they must retain their linkages with the operational processes, and with the object field of the knowledge which is to be developed. Hence, they cannot be transformed into isolate objects of instruction, as in Papy's books.

Naive set theory intended to be a means of understanding becomes an object of teaching.

The main reason for this is quite obviously that the concept of set is not a common concept that can be naively elaborated in a directly usable form. Is there, however, no possibility of simplifying exercises and some reasonings by the use of drawings?

"These round figures, or rather these spaces (for it does not matter which shape we give them) are very appropriate to facilitate our reflections about this matter, and to reveal to us all the mysteries which one boasts of in logic and which are proved there with much trouble, while with figures, everything jumps into the eyes." (Euler, Letters to a German Princess, 17.2.1761).

Hence, why not use these round figures in a corner of the blackboard or at the margin of a page to accompany the correct mathematical discourse (N. Bourbaki, 1951, pp. 12, 131,...164). The student, of course, will not require that the use of these figures be accounted for. Whether their function is essentially heuristic, as for Euler who uses them to enumerate the syllogisms and to assure himself of the completeness of his study, or, more simply just for show, as in G. Papy's first propositions, they are suggested as means, with the student being expected to make good use of them. "Graphs" are a simple means for teaching, among other things, the naive theory of sets (or the logic of classes) the value of which may crumble as soon as the pupils ask to see how it works.

The teachers who have to answer the pupils' questions, and respond to their failures, are not going to be able to retain naive set theory as the main object of teaching. In fact, they will be led, by the requirements of the didactical contract, and for the reasons we have quoted above, to transform the intuitive means of control into an object of teaching. Moreover, the means will have to function as a genuine model, the pupils rightfully requiring that they are able to answer all problems by means of the representations they have been taught. Now both Euler's round figures and Papy's "potatoes" do not permit this, and each contradiction will have to be solved by the conventions which must be included in the method.

To give an example: each complement of a set is a set. Now it must be represented as well, and hence also the universe of sets; while a figure designates in fact two sets which must be distinguished, be it by coloration, or be it by a notation of the boundary. The idea of "set" fits well to a set of points of the connected, compact plane ... the interplay of intersections, however, has done everything to produce parts that are in several pieces, which will be identified by coloring the boundary in the same color. It will be necessary to be able to indicate that one part is empty, and to distinguish a genuine list of elements (in which all elements are represented and identified, and there has been one author who used the term "complete" to the detriment of the analysts) from a drawing in which only a generic element or some special elements are represented. Even if one excludes drawings of the potatoes' "interior", the boundary determines in fact three regions, and not just one ... Every insufficiency, every incoherence of the metaphor can only be overcome at the cost of a convention, a specification, a complementary piece of information which teachers and editors require of textbook authors. The idea that everyone can and will solve these problems of representation individually does not excuse teaching from avoiding transforming all these conventions into objects of teaching, but rather adds to the confusion where the intention was to establish the simplicity and universality of one language. This in fact shows that the text alone (or any other form of explicit communication) cannot establish the metaphorical character of these presentations.

It is this other side of the fact that theoretical thinking has tacit elements that depend on personal contact. Instead, the spiral of the metadidactical nose-dive spins faster: we have seen that in order to control the use of Papygraphs that have become objects of teaching, conventions had to be developed which sharply increased the amount of metalanguage required: "potato, arrow, boucle ribambelle ("loop cluster") themselves acquire mathematical status, not only in the pupils' mouth, but even in certain volumes where definitions are formulated in terms like "Une relation reflexive est une relation bouclée partout".

Great efforts have been made to unify all these particular conventions and to develop a "didactics of graphs", but the step has not yet been accomplished. In the classrooms, however, the process continues between teacher and pupils: innovations of all sorts, whether introduced by the teacher or by the pupils, and which of course will be specific to the history of that class, are first objects of knowledge, then become objects of teaching, which have no cultural basis whatever and cannot be depersonalized.

The phenomenon which we have seen consists of transforming a means of education, or a means of understanding a piece of knowledge, into an object of instruction. It occurs most frequently when a difficulty or a failure has to be faced. In its most familiar form, for example, the teacher gives a first explanation to enable the pupil to acquire the knowledge, but the latter fails and (implicitly) holds the course responsible; while the teacher either changes the problem to make it simpler (Topaze effect) or looks for an analogous problem taking the first explanation as the object of the second explanation, hence as object of study, and finally as a subject matter to be taught.

The phenomenon of the "metadidactical nose-dive" just described has been going on for several years. This is all the more surprising as a large number of individuals who have received the best scientific training have been involved, and as there has been no lack of criticism and warning: it would seem to be irrepressible, and neither scientific alertness nor common sense seem sufficient to bring it under control.

That this is an effect of the contract established between teacher and pupil seems rather evident, but it is in fact a result of the fact that we live within a society entrenched with science and technology, without the average person being a scientist or having experienced the epistemology of research.

The example we have given clearly shows the role played by the fact that we are dealing with a fundamental and almost metamathematical theory.

Initially, the didactical contract leads to a slipping of functions between the knowledge whose transfer is intended, the means of controlling this knowledge (its understanding), object of teaching, and means of teaching. Later, or rather dialectically, the change of function profoundly modifies the elements present.

The requirements of the didactical contract lead the teacher to formulate as many means of control as possible, which the pupil may apply in using the knowledge the teacher intends to transfer while explaining the object of teaching, transforming these means of control, according to their needs, into pieces of knowledge and skills which are as text-determined and algorithmic as they can make them.

The condition of the didactical contract, the necessary and intuitive meta-understanding tends to be "methodized" and thereby to be destroyed.

5. PARADOXE SUR LE COMEDIEN (DIDEROT)

In school, the pupils have to "close" the problem, i.e. they have to answer a certain question and to present the answer in a certain desired manner. Schooling establishes above all else both the individual's conscious and unconscious relationships to knowing. "Indeed, like any social insight, the judgements that teachers make with regard to students, particularly in examination situations, take into account not only knowledge and know-how, but also the intangible nuances of manner and style which are the imperceptible and yet never unperceived manifestations of the individual's relationship to such knowledge and know-how ... (Bourdieu, 1977, p. 338).

But does style represent all aspects of the relationship to knowledge in question? To make a problem one's own, to have an "intrinsic" motivation, to get involved etc. etc., do all these seem to be activities in contradiction to the process and purpose of schooling, or not? The questions arising here are very brilliantly expressed in Diderot's "Paradox of Acting". "Les comediens font impression sur le public, non lorsqu'ils sont furieux, mais lorsqu'ils jouent bien la fureur" (190). And a few lines later Diderot writes: "... ce que la passion elle-même n'a pu faire, la passion bien imitée l'executé" (191).

What then does it mean to know? To be intrigued, engaged or occupied by a question or a problem? Is that enough, or is it at least the essential aspect? Or is it more essential to master the problems and skills in question? Do proficiency in the skills and ability to apply them correctly in mathematical problem solving represent the goals of mathematical instruction in school? The paradox of acting suggests something else.

The subject has to distinguish between oneself, the knowledge and the object, and only social interaction will help to achieve such an objectivisation, which makes realistic knowledge.

The paradox may thus be restated: in that movement only as my knowledge is no longer my individual personal involvement but is social knowledge, it is my knowledge. Even a person's most intimate experience is a conscious attitude only insofar as one's dealings with it involve explicit social knowledge. Only knowledge of a personal kind, however, can be true, because there is no such thing as consciousness as such or in general.

Diderot said: "... il y a trois modeles, l'homme de la nature, l'homme du poète, l'homme de l'acteur. Celui de la nature est moins grand que celui du poète, et celui-ci moins grand encore que celui du grand comédien, le plus exagéré de tous" (186). In Diderot's terms the actor is the greatest of them all because he himself is the object of his creative activity. For this reason he must have the power of self-elucidation. On the basis of this power his acting will be pure art having nothing in common with emotion, but making the greatest impression on the audience because of its authenticity.

The most dangerous property for an actor is "sensibilité naturelle" ... "c'est l'inégalité des acteurs qui jouent d'âme" (128) and: "Elle doit conduire peu a peu l'acteur a la manière et a la monotonie" (138).

As delicate responsiveness of the actor (or pupil) leads to monotony, so involvement and intrinsic motivation of the scientist (or pupil) leads to one-sidedness. Always "la facilité de connaître et de copier toutes natures" is very advantageous. It is the mark of the man without genius to be capable of only one thing, adhering to a single idea of truth, whereas the genius is "propre à trop des choses".

"L'homme du poète" in Diderot's essay is in a certain way and for the moment to be compared with the researcher. Theoretical knowledge, as produced by research activity, is personal, intuitive, one-sided. The process of objectivisation and substantiation of knowledge demands distinguishing, to a certain degree at least, between the "context of justification" and the "context of discovery". Discovery is what science is all about. Discoveries are made by genius. Yet the only evidence of genius is the success which the thinking produces. This success is not in the hand of the genius but is due to evaluation during the course of social history. The problem of discoveries can neither be reduced to the psychological nor to the objective level. This is taken into account by Diderot's conception of the actor as the greatest of genius.

But is it not strange that one should be able to characterize genius by certain properties, whereas their achievements cannot thus be characterized? In fact, Diderot had elsewhere, especially in his early works, proposed a conception of genius quite opposite to the one in the "Paradox of Acting", the essential features being enthusiasm, sensibility, one-sidedness caused by the predominance of one sense or of one organ etc. The genius is seen here as one who is carried away by the intensity of feelings.

It is evident that the notion of the genius as an individual endowed with such faculties is incompatible both with the standards of social or public success and with the idea of instruction based on the possibility of communicating everything to everybody.

Under the circumstances of the "didactical contract", one may even be led to the idea that to show talent is the most important thing. Understanding is identified with communicating or representing or producing. "Étre intelligent c'est faire ce qu'on attend de vous" was one of the accusations of the student's revolt in Paris in May 1968. The connection with the academic disciplines will not help here. The specialized researchers seem addicted to the same type of ideology, even though for quite different reasons (the ideology of the genius, that means the one being able to use the "rules of the game", the methods proper, in the most perfect manner). Enthusiasts or passionate people enact life's comedy, the true scholar or artist or pupil in school observes and studies them with icy detachment. The specific character of the "pupil as an actor" result from the fact that the relationships between life and school are complex and never clear nor stable.

Specialization and division of labour resulted in a distinction of the tactics and strategies of the coordination of the professional activities on the one hand and of the social-historical process of objectivisation of knowledge on the other. We also witness the historical evolution of new forms of interaction of these two contexts. To signify these tendencies, which render pure mathematics, as it appeared during the 19th century, a historically relative phenomenon, we have used the term "complementarity".

The assertion that "the mind is opposed to the objective world, because it is my way of treating the world, determined by my biography" remains true only insofar as I am not able to consider my biography as objective in the context of social history.

Diderot's paradox tells us something about the universality of the formal attitude. If "sensibilité naturelle" leads to monotony, the "formal approach" may result in objective flexibility and objective sensitiveness. Theory is among other things a way of seeing the world. The methodologism of specialized mathematical (or scientific) activity thereby produces the grounds for an universalism (mathematics as universal form) which may appear paradoxical sometimes.

Pure mathematics does not concede an epistemological role to the question of the application of knowledge, although the universal applicability of its methods and concepts is always stressed by pure mathematicians. Diderot, however, has told us that formal style is either mechanism (to best be implemented on a machine) or it is nothing but the representation of the non-formal.

The attempt to maintain meaningful links to the scientific is no longer possible, although it is a historical fact that the large reform projects during the '50s and '60s, which were committed to the scientific development of mathematics and other subjects, were also devoted to an interest of strengthening the commitment to the "didactical contract". The fundamental philosophy of these efforts is probably expressed strongest in Bruner's (1977) "The Process of Education". They failed because of a too naive and too optimistic view of the relationship between science and society.

6. THE "FORT-DA" GAME (S. FREUD)

The psychoanalytic theory about the origin and evolution of the symbolic function provides the social relationship of the subject with its mother with a fundamental importance, which the "fort-da" game allows to analyse.

The child makes a little puppet disappear and reappear again at its will and with a recognizable emotion. The separation from its mother, whose appearance or departure is neither foreseeable nor controlled by the child, causes frustrations for the latter.

The symbolic significance of the "fort-da" game for the frustrated child seems evident. One should keep in mind however that this significance is not constituted by the different elements – the puppet for instance is not a representation of the mother – but by the dynamics of the situation as a whole. What provides the situation with symbolic meaning is not its similarity to "reality" only, but the possibilities of bringing it into manifold and differently balanced relations with that reality.

An automatic puppet moving in a completely regular fashion cannot play the same role, as the child foresees the movement of the puppet and this destroys the child's interest. A puppet moving completely at random however would be too realistic and would only cause anguish for the child.

The continuance of the pleasure is linked to the disappearance of the puppet for the purpose of its renewed coming into view again.

Soon however the efficiency of the control over the puppet deprives the situation of its capacity to express emotionally the child's relation to its mother.

The purpose or function of the symbolic situation born out of the desire to compensate the frustrations and emotional tensions, deceives that desire through the very act of its apparent fulfillment. The symbol which was exacted to realize existentially that which was not realizable in the relationship with the mother inherits from the loss of that potential the original frustrations.

The creation of the symbol produces itself the "Kastrationsangst" (Freud) which in its turn makes it necessary to create a new symbol. The chain of meanings is without an end and is open.

Knowledge appears as the cause as well as the objective of the investment and involvement into a particular symbolic situation and also the cause of the disengagement and disenchantment.

For to provoke an initial investment into a situation, which shall lead to new knowledge, this situation should be open, uncertain, promising and should cause some uneasiness. It should allow for the play of desire and decisions, of hopes and anticipations, of resumption and repetition.

The situation should however concede a progressive control over the emotional investment only if the intended knowledge is attained. Then the situation gets closed. The quality of a person's cognitive behaviour is constituted by the quality of their previous knowledge and by the alternatives considered (and chosen or rejected) within the situation in question. With respect to knowledge, there are two movements inverse to each other which are asked from the subject in this process. First, the acceptance of the uncertainty and the risk of a game as well as the unprejudiced ignorance associated to those, and second the rejection of all that, confidence as well as the strong interest in the clarity and determination provided by insight and knowledge. The passionate player seeks this pleasure continuously and exclusively on the level of the first movement.

The serious study of probabilities requires, however, a distant and reflexive attitude opposed to the pleasures of playing.

In addition to that, it is not possible to have a prescribed or fixed hierarchical relation between the two movements, respectively the attitudes constituting them, nor to have one absorb the other. Only for LaPlace's "Universal Intelligence" are the two identical. But then there is neither knowledge nor uncertainty.

The human subject has in a certain sense to know what he does not yet know and has at the same time to forget some of what he already knows or believes to know.

One has to know more as well as less than one "knows". The acts of avoidance as typically shown by pupils illustrate this dialectic very well.

Altogether, one may in fact observe the inclination of certain pupils to take on willingly the first moment, namely the questions and problems, the uncertainty and opennes, the complexity and the playing on the verge of knowledge. But they cannot quit or leave this gambling, autotelic, unrewarded behaviour and cannot get involved in a really responsible manner with the second age.

For others, it is impossible to accept the first moment. Problems and open questions are not tolerable for them. Any uncertainty or openness seems to signify a state of primitive personal existence which they desire to have already overcome and surpassed. A question without an immediately conceivable answer causes anguish for them. They obsessionally ask the teacher for answers, for decision procedures to reach an answer, for algorithms.

There exists a second paradox in the way the subject develops his activities of symbolisation. To get involved in an open situation and to gradually gain control over it apparently represent, on the one hand exactly the conditions necessary for the acquisition of new insight and knowledge. On the other hand, one may consider these activities to be the usual employment of already acquired knowledge. As a person solves the problem inherent in the situation he obviously already has all the knowledge that is necessary for that purpose. The fact that he learned something from the situation then is manifested by his failure to solve the problem. The knowledge is the prerequisite as well as the result of the problem solving activity.

Especially the fundamental theoretical terms are, when one reads and tries to understand a text, for instance, at the same time the known as well as the unknown, the starting point as well as the objective of the subject's activity.

Knowledge or a theoretical concept is never to be exhaustively represented by any one problem, activity, text, object field etc. (We encountered this phenomenon already on the epistemological level: the paradox of theoretisation). This paradox places learning into the comprehensive context of the subject's objectified biography. And for the activity of historically objectifying one's own personal biography knowledge contributes a fugitive, elusive but centrally essential element.

The ideal of knowledge is an identity of being and knowing or thinking. The cartesian "I think therefore I am" has therefore to be counter-balanced by an attitude which founds the objectivity of knowing on the real being. (As the real being is not a symbolic process, this gives hints to the potential of symbolization).

From the history of mathematics we know many instances where it was not the referential meaning of a sign which was essential but where a symbolization like the above could only be understood when related to the dynamics of the problem situation as a whole. On the other hand we know that the dynamics of the problem situation are intimately and inseparably linked with symbolization.

Not only does solving a problem often mean to represent it so as to make the solution "obvious" or transparent. Without seeing symbolization linked to the dynamics of the situation, we would not be able to understand how somebody could find the answer to a problem unless he already knew the answer (cf. Plato).

Problems (intentions, goals) and symbolic means and models are the two sides of one coin. Empirical research presents findings saying that students with higher mathematical abilities, when trying to solve word problems by means of equations, did not – as the less able students did – try to name all the occurring magnitudes or unknowns, but instead immediately looked for the relevant relations or equational relationships (cf. Doblaev, 1969).

7. CONCLUSION

The "infinite intelligence" of LaPlace can know everything about the world under the condition of being exterior to this world, and, as a consequence, it does not need to conceive it theoretically or historically. Against the geometrized and static world view that underlies the traditional epistemologies of mathematics we should like to promote the belief that "time" should as an essential element be introduced into all our reflections on mathematics and mathematics education.

Each of the sections in this chapter attempts to demonstrate a different aspect of the fact that the human being, on the contrary, is at the same time both the subject and the task of cognition or the source and the object of activity.

This twofold necessity leads to the apparition of a whole series of pairs of concepts: insight and action, intuition and formalism, knowledge and application, and so forth; the character of which we have tried to show as being both paradoxical and necessary, opposite and complementary. These oppositions are the source of the fragility of the act of knowing and the difficulties in the transmission of knowledge.

Our remarks certainly appear to be somewhat negative and very distant from the preoccupations of teacher trainers who are essentially oriented towards practical action. Is it not after all true that despite their alleged fragility, knowledge and insight manage to exist, develop, function, spread, be useful, and increase? Perhaps we have only emphasized the weakness of our knowledge about learning and knowing instead of the fragility of knowledge itself. Perhaps we have rationalized a checkmate in our attempt to theorize knowledge and its communication?

Some could argue that the reactions of the system to attempted modifications and reforms are the proof of the solidity of the mechanisms of control: They confirm the conditions of the communication of knowledge, and through this, are a witness to their strength. Moreover, what part of social activity is really threatened by this alleged fragility of knowledge.

In fact, the repeated attempts at reforms are, on the contrary, the proof of the dissatisfaction that results from the actual division between theory and practice, between work and learning. Education prepares the students to accept a social structure in which some are given the algorithms without meaning, and others the awareness without its products. Is it necessary to accept the integration of the corresponding epistemology in the curriculum of general education? Do we also have to plan for two types of person? And if not, to what extent does a better educational curriculum depend on a better understanding of didactic and cognitive processes?

It is clear that the significations constantly change and disappear, other ones that are similar or related are unceasingly generated by the same kind of social conditions and relations. Is this the true source of the generations and changes of meaning? In this case, we would have to envisage three sorts of educational action that are interrelated:

1. An action on the relationship between the teacher and the student that would provide situations permitting a much improved simulation of the operation of learning.
2. An action on the relationship between the teacher and the society based on the necessary epistemological conceptions that are held in common by this society.
3. An action on the structures by which this society claims to control the diffusion of knowledge and its uses.

REFERENCES

Bateson, G.: 1973, *Steps to an Ecology of Mind*, Paladin, London.

Bourbaki, N.: 1951, *Topologie Generale*, Hermann, Paris, Second edition, Chapters 1 and 2.

Bourdieu, P.: 1977, *Outline of a Theory of Practice*, Cambridge University Press, Cambridge.

Bruner, J.S.: 1977, *The Process of Education*, Harvard University Press, Cambridge, Mass.

Cantor, G.: 1932, *Gesammelte Abhandlungen*, Springer, Heidelberg.

Doblaev, L.P.: 1969, in Kilpatrick, J. (ed.), *Soviet Studies in the Psychology of Learning and Teaching Mathematics*, SMSG, Yale.

Euler, L.: 1761, *Letters to a German Princess*, Part 2, Letter XXXV.

Godement, R.: 1963, *Cours d'Algebre*, Hermann, Paris.

Howson, A.G.: 1973, *Developments in Mathematical Education*, Cambridge University Press, Cambridge.

Jakobson, R. and Halle, M.: 1971, *Fundamentals of Language*, Mouton, The Hague.

Popper, K.R.: 1972, *Objective Knowledge*, Oxford University Press, Oxford.

Schwartz, J.: 1983, *Tyranny, Discipline, Freedom and License: Some Thoughts on Educational Ideology and Computers*, M.I.T. Press, Cambridge.

Wittgenstein, L.: 1967, *Remarks on the Foundations of Mathematics*, Oxford University Press, Oxford.

Wooton, W.: 1965, SMSG, *The Making of a Curriculum*, Yale University Press, New Haven.

ENTR'ACTE 2

The awareness of paradoxes is taken up in a more general way by Mellin-Olsen in the next chapter. Where Otte and Brousseau were concerned with the fragile nature of mathematical knowledge in the teaching/learning situation, Mellin-Olsen explores the double-bind trap which always waits for any teacher, and the mathematics teacher is no exception. The double-bind in general occurs in any situation involving double communications which contradict each other. In particular the author employs Gregory Bateson's double-bind structure as a tool for didactical analysis, especially in the context of goal-directed activities. He is particularly concerned with the possibility of teachers producing damaging double-binds by transforming a goal-directed learning activity into a purposeless learning activity. As it is hard for the mathematics teacher to detect double-binds with her students, this chapter's analysis enables the teacher and the teacher educator to avoid the production of new double-binds rather than to detect and dissolve existing ones.

STIEG MELLIN-OLSEN

THE DOUBLE BIND AS A DIDACTICAL TRAP

1. INTRODUCTION

Bateson (1973) introduced the notion of a double bind in the context of mother-child relations in various cultures. The double bind refers to a double communication from mother to child in which the two modes of communication contradict each other. For instance could the verbal expressions of the mother express attraction at the same time as her facial gestures express rejection?

The damaging effect of such a contradicting communication occurs when the individual who receives the messages does not discover the conflict. The individual is then bound to two contradicting modes of communication, and is thus prevented from taking any rational action in relation to the messages of the communication.

The result may be behaviour which is irrational to the observer: aggression, passiveness, confusion, lack of creativity, etc.

Diderot's paradox of acting reported in the article by Brousseau and Otte (pp. 29–32, this volume) illustrates the classic double bind. Diderot's advice to the actor is to act a certain emotion, rather than to communicate it through his own emotional system. The actor's audience is thus not to relate to the actor as an individual but to his acting.

We can compare the teacher with the actor and the students with the audience. Diderot's advice then transforms into advice to the teacher: The students should relate to the knowledge communicated to them by their teacher rather than to the teacher as an individual.

In this case the double bind occurs when the pupils are not able to distinguish between the teacher's expectations of them and the possibilities of knowledge. The teacher now does not manage to release knowledge from her own person. The double bind becomes hidden in the double message from the teacher to the students: "I want you to understand this knowledge, but in order to do so, you have to understand me at the same time". Or: "If you reject this knowledge, you also reject me." The latter message prevents the students from accepting the teacher at the same time as they reject her knowledge.

The double bind structure as it was developed by Bateson and his followers, such as the British psychiatrists R. Laing and A. Esterson, can be generalised in various ways. In its most general form the double bind

Alan J. Bishop et al. (eds.), Mathematical Knowledge: Its Growth Through Teaching, 39–59.

structure can be considered as an explanation of the socialisation of the individual to her society (Mellin-Olsen, 1987). At this level the individual takes the basic norms and values of society for granted. No critical questions are asked. The individual is not aware of possible contradictions between her own life experiences and how her social environment teaches her to code these experiences.

The double bind as a tool for socialisation is hidden in messages to the child/student such as

- you should be creative but not too creative
- you should take control, but not control too much
- you should take care of your own interests but don't forget other people
- you can do it your own way as long as you are not breaking the rules.

The individual is anchored to her society through norms and values. It is such anchors which leads Jules Henry to claim that if US schools should take their goal to educate creative citizens seriously, schools should start by teaching the students to doubt the Ten Commandments and the 2 party system.

Elsewhere I have studied how the double bind structure can be used as a tool to understand many teachers' resistance to changing their understanding of learning and teaching (Mellin-Olsen, 1988). Taking Marton's (1988) position that learning is related to a shift in the understanding of a phenomenon, the double bind structure can be used to explain cases where an individual apparently is not able to change such an understanding.

Marton suggests that changes between understandings of a phenomenon constitute the most important form of learning. He claims it is necessary to specify the kind of learning and understanding the teaching methods are aimed at bringing about.

Applying this line of reasoning to (teacher educator-student/teacher) relationships we can say that Marton's paradigm guides the teacher educator to convey certain understandings of various phenomena related to learning and teaching. As a result of their lived experiences student teachers have some understandings of what learning and teaching are about. According to Marton's paradigm the teacher educator should eventually bring about changes in such understandings for the purpose of developing of new qualities in learning and teaching.

The double bind structure demonstrates that sometimes it is very difficult to bring about such changes. Those students who by their own learning history have a certain understanding of learning and teaching which contradicts the rationality of their own experiences, react differently to a teacher education which aims at changing their understanding.

Those students teachers who have insights into the contradictions embodied in their learning history can, if it feels right, welcome intended shifts in their understanding of teaching and learning. Those who do not have such insights into the contradictions may have severe difficulties in adopting new understandings, that is, in changing the patterns of their own learning history.

Remembering that a double bind is a basic structure related to the deep adaptation of the individual to her society, we see that a quest for a changed understanding of a phenomenon may threaten the individual in such a way that she may even hang more tightly on to her present understanding.

Marton's paradigm about learning as a shift in understandings of a certain phenomenon can thus be controlled by double binds. Double binds may prevent certain shifts, thus surprising certain learning possibilities.

In Mellin-Olsen (1989) I have analysed how student teachers react differently when the teacher educator provokes some double binds related to certain understandings of learning and teaching.

In the following I shall analyse certain ways a teacher can produce certain double binds with her students. In the shadows of this elaboration lives an assumption that as teachers themselves are victims of certain double binds they will communicate contradictory messages to their students.

2. THE DOUBLE BIND STRUCTURE

A double bind occurs in an individual as a result of contradictions in communication. Bateson (1972) is not consistent in his uses of the concept. Several characteristics of the conceptual structure can be listed and Bateson stresses them in various ways in different papers. The most important properties of a double bind are:

A double bind is a state within the individual as a result of messages sent to her in a communication system.
- The messages communicated to the individual contain a contradiction.
- The individual does not disclose the contradiction.
- The individual has no opportunity to escape the communication system by, for instance, turning her back on it.
- The communication is vital for the lived life of the individual.
- The communication situation is repeated.

The contradictory message in a communication can be conveyed to an individual in several ways. In his communication theory Bateson distinguishes between knowledge and metaknowledge. He similarly distinguishes between communication and metacommunication, and knowledge (level 1) and metaknowledge (level 2) (see Brousseau and Otte, pp. 21–22, this volume).

Furthermore he distinguishes between two basic forms of communication: that is, analogue communication and digital communication. Verbal language is the classic example of a system which produces digital communication. The communication accompanying verbal language, such as uses of voice and facial gestures – the paralinguistics and kinesics to use

Bateson's notions – is analogue communication.
Tools of analogue communication are
- facial gestures
- loudness of speech
- a violin

Tools of digital communication are
- words and sentences as in spoken language
- a piano

Double binds are produced in the classroom almost every lesson. Take the Bert case reported in the Introduction. The giant Bert's hand is drawn on the blackboard as a means to introduce ratios.

> T: How big is the giant? Many children join the discussion giving their opinions: "As big as the classroom" ... "as big as a tree". (But these answers the teacher didn't ask for.)
>
> Look at my hand and look at the giant's hand. Compare both! (She puts her hand beside the big one, on the same line.)

How do the pupils understand this didactical situation? Many of them will probably understand that they should seek the teacher's understanding of ratio rather than their own. As soon as the pupils start to explore the mathematics of the situation according to their own interpretation of the situation, the teacher immediately stops them.

This does not necessarily lead to a double bind. The risk of producing a double bind occurs if the teacher already has conveyed an expectation that knowledge should be personal:

"You should learn to use knowledge in your way."

"You must try to see how you can use what you know when solving problems."

"Please – try to think about what you can before you ask me about how to do."

The contradiction is here located in the polarity of teacher's knowledge/ pupil's knowledge. The teacher conveys knowledge as it belongs to her: "But these answers the teacher didn't ask for." At the same time she may communicate that the pupils should take on board knowledge for the purpose of individualised use.

Similar contradictions in communications may be found in the other cases in the introduction:

> Immediately one begins to wonder how the pupils understand this activity – is it a game? What are the rules? What is the right answer? Indeed, what does it mean that a pupil understands an activity? What do we mean when we talk about the understanding of rules? And what of this teacher's role? What goals does he have? *He* seems to know because he evaluates the

pupils so quickly. But should the teacher tell the pupils what the goals are?

New problems emerge in relation to the double bind structure:

> What significance does it have for the pupils' learning if they do not know the teacher's goal? We assume that they must have some goal for their learning. This goal may be different from the teacher's. It may be similar. What happens if the teacher persistently moves in another direction with knowledge than that which the pupils does?

It is possible to extend the double bind structure to see how an institution can communicate contradictory messages to its individuals. A well known example is the grading systems in schools which are based on an ordering of students according to percentiles. Such a system implies that a given proportion of the students will achieve the lowest grade. When the messages to the students of such a system accompany messages which talk about equal opportunities a double bind can easily occur. This will be a double bind produced by teachers who are loyal to the logica of the rules of the institution (Howson and Mellin-Olsen, 1984; Mellin-Olsen, 1987).

3. CONTROL OF KNOWLEDGE

I shall disclose some classroom situations which are vulnerable to the communication of double binds to pupils. In order to do this I need a didactical structure within which I can locate the double binds. I choose a structure which is based on the relation between the individual's activities and her control of knowledge (Mellin-Olsen, 1987a, 1989).

To the notion of "control of knowledge" I relate two features:

(i) I relate *independence* to the concept of control. I see a pupil control some knowledge if she can use it independently of the teacher. This is, as I interpret it, in the kernel of the didactical contract as described by Brousseau and Otte (see Brousseau and Otte, pp. 18–22, this volume). When the pupil breaks the didactical contract she is not dependent on the teacher in her use of the knowledge any more. She can go her own ways with the knowledge.

(ii) I relate an aspect of metaknowledge in relation to control of knowledge. I find it necessary to stress that if a pupil shall control some knowledge, she needs to be aware of the availability of such control. This implies that the pupil should understand herself to be someone who can master knowledge in the field she works within and thus has the opportunity to exercise knowledge control.

The non-triviality of this statement is revealed whenever a teacher is convinced that her pupils master some knowledge to perfection, and still becomes surprised when the very same pupils fail to employ the knowledge in some problem solving activity.

I see three possible levels at which a pupil and teacher can negotiate about knowledge control:

N1. Tool Level

Control of uses of knowledge as a tool when working with a problem.

In this case the pupil is expected to work with a problem and the teacher has provided her with suitable tools. The situation is thus: Who is to control the uses of the tools, the teacher or the pupil?

N2. Choice Level

Control of choice of knowledge as tools when working with a problem.

In this case the pupil is expected to work with a problem. Who is to control the choice of suitable tools to work with? The teacher or the pupils?

This level includes the situation where an appropriate tool has to be constructed. Who controls the construction of the tool – the teacher or the pupil?

N3. Goal level

Control of goals of problem solving.

Who is to control the goals of a problem solving activity? The teacher or the pupils? Who is to control the development of the problem solving project as it proceeds?

We dichotomise the outcome space of control possibilities by giving either Teacher (*T*) or Pupil (*P*) control at each level. For simplicity I thus disregard the possibility of a dialogue situation or a negotiated situation. We end up with 8 possibilities:

Level	N1	N2	N3
1	*T*	*T*	*T*
2	*P*	*T*	*T*
3	*P*	*P*	*T*
4	*T*	*P*	*T*
5	*P*	*P*	*P*
6	*T*	*P*	*P*
7	*T*	*T*	*P*
8	*P*	*T*	*P*

This simplistic model provides an opportunity to analyse the dynamics of a didactical situation:

Example 1

$(T\,T\,T) \rightarrow (T\,T\,P)$

This development shows us that the teacher originally possesses control at all levels. This results in a situation where the pupils have grasped control at the goal level.

Example 2

$(P\,T\,T) \rightarrow (P\,T\,T)$

This development shows that the teacher first possesses control of the choice of tools which are to be used in the problem solving process. Furthermore the teacher has control over the goal of the project. In the second situation the pupil has grasped control at the choice level (N2) but the teacher has regained control of the goal level.

We get a preliminary idea about the possibilities of damaging double binds:

The teacher may communicate a desired didactical development such as those illustrated in the examples above. During the didactical processes she may, due to her own double binds or due to lack of didactical insight or to a combination of these factors, communicate a message which contradict the first. This happens when she, for example

conveys the message: you are to take control

at the same time as she

conveys messages which prevent pupil control.

4. TWO PROJECTS

I shall analyse two classroom examples to demonstrate the argument. I choose them from project work where the mathematical tools are used on problems which are of a non-mathematical nature. I base my analysis on texts written by teachers. The texts I analyse are extracts from project reports as part of a postgraduate examination in mathematics at Bergen College of Education.

The two reports are selected from a sample of reports which I have analysed elsewhere. The project was to see how teachers take care of pupils' knowledge interests when they are asked to do so (Mellin-Olsen, 1989). I have selected the two texts which gave the best evidence about how teachers can succeed in letting their pupils grasp control.

Later I shall examine some weaknesses of even these texts in order to demonstrate some strengths of the double bind structure as a tool for didactical analysis.[2]

P1. Road Safety. 10 Year Olds. Rural Area.
Extract from Teacher's Project Report

I introduced the problem to the class:
Where are the dangerous spots on your way to school? What can be done about them?
...
Almost everyone mentioned a narrow and blind bend where several incidents had happened already. After the discussion each pupil drew a map of her school road. They picked the places which appeared most dangerous. That was the beforementioned bend, a narrow bridge and the crossing of a heavy used main road.

COMMENT

A map is to be drawn. Who controls the drawing of the map? The map is to function as a tool for further investigations. A tool has to be constructed.

Questions: Do the pupils know which function the map has in the ongoing learning activity? To what extent does the teacher interfere with their construction of the map? Are certain formal rules for map construction introduced here which disturbs the development of the project?

Who control the drawing of the map – the pupils or the teacher?

The pupils were also interested in counting the cars on the main road. This should be carried out in the morning because most of the pupils would be coming along the road by then.

After the investigation that morning we went back to school together. As we walked along the pupils commented in an animated way their counting. One of the pupils wanted to do the same investigation in the evening when the pupils were on their way home from school. That was not possible for practical reasons, but it was nice to hear him say this.

COMMENT

Why was it important for this pupil to count in the evening? Just for fun? Or was there an idea behind the suggestion?

Why was it not possible to do the same investigation in the evening? Was any reason given to the pupil? Did the pupil accept the reason? Was any solution to this practical problem discussed?

In the classroom the data were counted. We wrote the results on the blackboard and systematised them under five headings.

The pupils were all engaged in these calculations. One commented that had they counted an hour before, the figures would have been much higher because of the morning rush.

Some of the pupils checked this later.

COMMENT

What happened to the results of this counting? Why did the teacher not include these results in her report? Did she discuss these statistics with the pupils? Or did she only forget them?

Some of the pupils decided to question the secondary school pupils about why they crossed the main road rather than using the tunnel. Three different types of answers were reported:

- It is faster to cross the road.
- so many bicycles are parked in the tunnel.
- don't know.

COMMENT

Why did the teacher not report the pupils' discussions of this result? Were there no discussions? If not, why not? Had the pupils finished their participation in the project at this stage?

The original plan was that the pupils should present part of their work at a traffic day arranged by the community's welfare society. On this day both children and their parents met, and several issues for various age groups were discussed. The pupils were to present statistics about the following issues:

1. Dangerous spots on school road.
2. Time spent by pupils in traffic
3. Pupil age when they got their first bicycle

COMMENT

"The original plan". Whose plan was this? How did the pupils control the content and form of their presentation? What happened during the meeting? Did this project about road safety result in any practical actions from the authorities?

P2. Equal Rights for Both Sexes. 15 Years Olds.
Urban Area. Extract from Teacher's Project Report

I presented the idea of a project about equal rights to the pupils. They themselves wanted to split into groups to discuss what they meant about sex roles in today's Norway and if there are equal rights in Norway today.

Lively discussions.

"We must have more places in kindergartens so children can get into them," one of the most eager girls said.

"What is the relevance of that to equal rights" asked a boy.

"Don't you understand. It is difficult for the girls to work if they have children. The husbands don't want to stay at home. That is why this has to do with equal rights."

COMMENT

The teacher presents an idea of a project. She thus possesses control at the goal level. Her approach is, however, so open, that the pupils immediately grasp the opportunity to discuss. This stage of the project is significant as it provides an opportunity for the pupils to take control over the goal of the project. Key words in the discussions as reported in the text above are

kindergartens
relevance of equal rights
responsibility for children.

These discussions apparently do not influence the next stage of the project directly. Still the pupils – according to the text – seem to possess initiative: ...

... The group discussions resulted in posters which summarised the discussions;

1. Same pay for same work.
2. Equal educational rights.
3. Same teaching and same treatment by teacher in class.
4. Everyone has equal right to express opinion.
5. Equal rights in home.

6. Women on the throne.
7. Same sports for both sexes.
8. More women in all sorts of administration and bureaucracy.
9. Women must be more clever to take responsibility.
10. Women in the army (national service).

After this the pupils wanted to interview people to listen to their meaning about some of the issues.

One group decided to interview people about education, professions and the work place.

One group chose sport, interests and pocketmoney.

One group wanted to examine people's opinions about who is to decide in society.

One group was to examine distribution of work at home, that is domestic activities.

The groups worked out questionnaires which were presented to the teacher. Even if they are the proposals of the pupils, the teacher should see them since the pupils are to use them out of school.

COMMENT

"After this the pupils wanted to interview people about..."

What is the purpose of this project?
What is its goal?
Information? Pupils' to become wiser about equal rights? Moving barriers?

Have the pupils really grasped control over the project? To what purpose? The text does not tell us.

"The teacher should see them since ..."
Very few would criticise the teacher for such censorship. Still – if the pupils are challenged to grasp control over a project such actions of the teacher are problematic. How did the pupils react to it?

After interviewing, the pupils worked at the data they had collected. Some discussions on procedures arose.

The pupils developed posters for presentation in class. They asked these questions:

– What can we tell from the data we have collected?
– What do the interviews tell us?

COMMENT

What was the result of the initiative of the pupils in this project?

The diagram showed clearly that more men were employed in each age group.

The pupils could also read from the diagram that the difference between employed men and women was largest between 25 and 40 years and above 50 years. Could the pupils explain this?

"That is when we get children, and Mum has to stay home as it is difficult to get someone to look after the children".

"Why can't the man stay home then?"

Boys protesting wildly.

"And also – it is the man who has the best income" one of them said. "Then he obviously has to be the one employed."

That was a statement we also wanted to examine further.

COMMENT

The report never showed how this statement was further examined.

Analysis

In both these projects the teacher introduces a project theme. There is a goal related to each:

To do something with road safety.

To do something with equal rights for both sexes.

There is a metacommunication: The pupils are expected to use their expert knowledge (as in the case with road safety) and their own knowledge interests (as in the case with equal rights).

At the initial stage of the projects the teacher possesses control at the goal level:

$$(* * T)$$

In both projects the pupils reacted positively to the idea of the project. There were several signs of how they tried to grasp control at goal level. They discussed how to proceed and they chose sub goals and methods to achieve these.

A new didactical situation thus emerges:

$$(* * T) \rightarrow (* * P).$$

Sometimes the pupils chose tools:

(i) they wanted to interview
(ii) they wanted to count a phenomenon
sometimes the teacher chose:
(iii) each pupil drew a map.
This leads to the possibilities
(i) and (ii): $(* \, T \, T) \rightarrow (* \, P \, P)$.
eventually leading to $(P \, P \, P)$
(iii): $(* \, T \, T) \rightarrow (* \, T \, P)$
eventually leading to $(P \, T \, P)$.

Both these developments can be productive. The difficulties arise when there is a mismatch between the messages conveyed to the pupils at each level:

You have expert knowledge
You are to follow your knowledge interests
It is your project
How can we promote changes as a result of our investigations?

followed by messages at level 1 or 2 which may contradict:

Don't use that method, use that one!
You should not investigate this issue, you should rather go for that one.

The two texts reported here tell us nothing about control at the tool level. They describe several significant incidents at the goal level:

- The pupils were also interested in counting the cars on the main road.
- As we walked along the pupils commented in an animated way on their counting.
- They themselves wanted to split into groups to discuss what they meant about sex roles ...
- The group discussions resulted in posters ...
- After this the pupils wanted to interview people ...

Such statements can be taken as evidence of how pupils attempt to grasp control over a project. These incidents are significant in a didactical perspective as the teacher here can either block the pupils' initiative or she can support them.

The teachers' sensitivity to her pupils' motives and her position as a listener are vital here.

Here the double bind works when the pupils are blocked within a frame of support without discovering the contradiction. In this case the teacher, as a subject or as an agent of the school's intrinsic control system exercises an efficient method of silencing the classroom.

5. SCAFFOLDING LEARNING AND THE DOUBLE BIND

The texts reported above and the comments may provide evidence about an already existing double bind among the teachers and the pupils. When they are invited to grasp control over the project they often do not know how to handle this opportunity. Neither does the teacher. Both texts include descriptions of learning activities. A phenomenon is counted. It is related to other phenomena. There are discussions, agreements and disagreements about interpretations.

The texts raise, however, so many questions which are not answered. They furthermore describe results without stating what these results could imply for further action. In short, the texts report so much purposeless activity.

This raises the hypothesis that the production of many double binds is related to neglect of the goal of the learning activity.

The texts reported above call for questions such as:

To what extent is the teacher aware of the goals of the learning activities during the project work?

To what extent are the pupils aware of the goals?

To what extent is the goal level forgotten, such that goal directed activity in class is reduced to just – activity?

So many of the comments on the texts are of the type:

- For what purpose?
- What did the pupils' findings lead to?

When the goal level of control, N3, disappears as a directing frame of the learning activities, double binds can emerge quickly. An understanding of what is going on such as

- The original idea was road safety. But it is all right to count cars, it is nice to go out.

develops to

- We have been busy for days now counting cars. It would be nice to work indoors again.

In order to combat such developments I shall suggest a didactical structure which forces the educator to be aware of the goal of the learning activity. That is scaffolding of learning. Scaffolding learning can be considered as complementary to double binding. As scaffolding promotes independence double binds produce dependence. Scaffolding promotes and supports the pupil's action and control. Double binds promote mock cooperation and purposeless reproduction of knowledge.

Vygotsky (1978) introduces a didactical structure which he calls *the zone of proximal development* for the child. This zone illuminates the child's potential for learning. Vygotsky thus distinguishes between the set of problems which the child can master without help from an adult and the set of problems which the child can master with support. The latter set of problems defines potential for learning. Vygotsky thus distinguishes between the child's actual level of problem solving and her potential level.

The proximal zone of development is thus

... the distance between the actual developmental level as determined by independent problem solving and the level of potential development as determined through problem solving under adult guidance or in collaboration with more capable peers (Vygotsky, 1978, p. 86).

In the perspective of this definition Vygotsky criticises the trend in education (of his time – the 1920s!) to evaluate learning in terms of which tasks the pupil masters rather than which terms she has a potential to master.

As I interpret this didactical structure I find that it calls for a stronger belief in pupils' learning capacities than we usually observe with teachers and curriculum-developers. The educator's conception of the pupil's learning potential may be radically changed if she looks for what sort of problems the pupil can solve under guidance, rather than if she plans the pupil's further learning according to what she already masters.

Complementary to this didactical structure, Wood *et al.* (1976) develop the didactical structure of *scaffolding*. This structure corresponds to the guidance called for in the definition of the zone. The metaphor "scaffolding" is borrowed from the handicraft of building: Scaffolding a house, for instance, implies a support system which works until the house stands by itself.

Bruner (1985) directly relates this structure to Vygotsky's zone structure. In this work he misses out one central issue which is basic for Vygotsky. That is the significance of the goal and motive of the child's problem solving activity. For Vygotsky the child's problem solving activity is related to a goal which should be – if not set by the child – shared by the child. In fact, Vygotsky in his writings is so consistent in his uses of "activity" with respect to this issue that when his pupil A.N. Leont'ev developed his grand activity theory, he could build directly on Vygotsky's more implicit uses of the concept.

Bruner (*ibid.*), however, is indifferent in relation to the significance of the goal of a learning activity. He refers to the experiments reported in Wood, Ross, Bruner (1976) and explains that the child initiates the problem solving activity because – as he expresses it – children usually do what they are asked to do by adults.

In another work of his the goal set by the child is ultimately present (Bruner, 1983). In this work Bruner constructs a conceptual framework which he calls the *language acquisition support system (LASS).* LASS functions as a scaffolding support in relation to the development of speech with the child. In the examples Bruner provides in this book the support for the child's language acquisition is related to a goal set by the child.

Greenfield (1986) has developed the scaffolding structure further. For her the presence of the learner's goal is a necessary constituent of the didactical situation. Greenfield describes the scaffolding as a support system which "...provides a support; it functions as a tool; it extends the range of the worker; it allows the worker to accomplish a task not otherwise possible; and it is used selectively to aid the worker when needed" (*Ibid.* p. 118). In her analysis Greenfield stresses the difference between Skinner's structure of *shaping* and the scaffolding process. In the shaping process the problem which the pupil faces is transformed, such that she can solve it by means of operations which she already masters. In this case we can say that the support to the child implies a change in the goal of the problem.

Contrary to this the scaffolding process leaves the goal of the problem unchanged. The support is directed towards the operations which lead to the goal.

This principle comes clearly through in Greenfield's example about girls who learn to weave in a Mexican handicraft culture. The skilled female weavers help the novices through the difficult parts. When the girls have completed the product, an expert eye was required to see that the product had not been made by a skilled weaver.

In the weaving example both the teachers and the learners control the scaffolding as a tool. Sometimes a girl addresses her teacher for help and sometimes the teacher interferes in order to help. I believe that sharing control in this sense is an efficient strategy as long as it is in underlying assumption that the ultimate goal is the pupil's independence in relation to the teacher.

In Greenfield's weaving example the role of independence is obvious: It is a shared goal that the girls should be able to weave with a certain skill when they are on their own. This role of independence is thus directly related to the scaffolding process: The scaffolding functions as a tool for the learner to achieve independence in relation to other people within a certain field of goal directed activities.[3]

This function of the scaffolding, to promote independence, produces another criterion, formulated by Lindén (1989): An essential part of the scaffolding should be that its application makes itself superfluous. In the weaving example we can expect the expert weavers to structure their support in order to reduce the risk that help will be required next time the same sort of trouble comes up. This dependence/independence dimension calls for a recognition of what we might call "false scaffolding". A false scaffolding is a support system which carries with it all the properties of the scaffolding, except for its mechanism to develop the pupil's independence in relation to the teacher. Such a false scaffolding automatically brings forth double binds:

> The pupil can understand that she is helped to be independent at the same time as she does not experience this help.

The growing independence of the learner as a result of scaffolding must be related to the goal of the learning activity. In order to get hold of the independence offered by the scaffolding, the learner has to be aware of the function of the scaffolding: What is the aim of the support system she is offered? If the pupil shall maintain control of her own learning activity she should at least get support to see these activities in the perspective of some goal.

Returning to the reported texts and my comments on them, we can see how the initial goal of the project is not always present in the learning activities.

P1. Road Safety

COMMENT

A map is to be drawn. Who controlled the drawing of the map? The map is to function as a tool for further investigations. A tool has to be constructed.

Questions: Do the pupils know which function the map has in the ongoing learning activity?...

Possible messages conveyed to the pupils:
- This is a project about road safety. A map has to be drawn
- Draw the map as you like it

or
- Draw the map in the following way: ...

What is the function of the map drawing in the context of the goals of the project? The text does not tell us.

COMMENT

Why was it important for this pupil to count in the evening? Just for fun? Or was there an idea behind the suggestion?
Why was it not possible to do the same investigation in the evening? Was any reason given to the pupil?

This is an example of independent action from the pupil which the teacher blocks. The text does not give any explanation.

COMMENT

What happened to the results of this counting? Why did the teacher not include these results in her report? Did she discuss these statistics with the pupils? Or did she only forget them?

Scaffolding has been provided by the teacher by helping her pupils to carry on a project about road safety. The text reporting this work, however, rarely informs us about any implications of the pupils' work. We see the risk of contradicting messages:

It is important what we are doing
b u t
we do not use what we find.

The same possibility emerges from the following comment:

COMMENT

Why did the teacher not report the pupils' discussions of this result? Were there no discussions? If not, why not? Had the pupils finished their participation in the project at this stage?

P2. Equal Rights for Both Sexes

COMMENT

After this the pupils wanted to interview people about ...
What is the purpose of this project?
What is its goal?

Initially the teacher supported her pupils when they prepared interviews. Her support nicely matches the requirements set by the definition of the scaffolding. Having done this the function of the interviews is not reported.

COMMENT

What was the result of the initiative of the pupils in this project?

The two texts raise the following question:

To what extent were the projects transformed from a goal-directed activity into a purposeless activity?

Such a transformation can generate double binds. The basic contradiction located in the question is the original presence of a goal which gradually shades off during the project. The seriousness of such double binds grows when this contradiction recurrently turns up in the didactical situations.

An example of this which can frequently be observed is the teacher who starts the lesson by motivating her pupils for a new field of knowledge, and then conveys the knowledge independently from the context of motivation originally given. The teacher's motivation then functions as a methodological ritual rather than as an organic part of the learning process as a whole.

The production of double binds can grow when the interrelationships between the various stages in a learning process are neglected. Then messages conveyed to the pupils at one stage may easily come to contradict messages conveyed at another stage.

6. SOME DIDACTICAL IMPLICATIONS

Theories which attempt to explain childrens' learning difficulties in mathematics have usually been based on the idea of their conceptual deficits. A growing area of research has limited theories which focus on communication. Both Sinclair (1984) and Johnsen Høines (1987) warn against the tendency to explain learning difficulties in mathematics solely by referring to poorly developed concepts. They argue about the importance of considering contradictions between how children and schools represent mathematical knowledge, stating that childrens' problems with learning mathematics can be due to communication problems rather than to lack of concepts. Advocators of ethnomathematics, such as d'Ambrosio (1985) and Gerdes (1986), generalise this paradigm by pointing to the destructive effects of the cultural alienation mathematics education might imply.

Double bind theory offers an alternative explanation of failure to learn mathematics. It can be profitable for a teacher of special education to search for possible double binds with her pupils in order to detect them and dissolve them.

A teacher who daily faces large groups of pupils can hardly develop such strategies. Rather what she can do is to combat the production of new double binds. She can also work against a reinforcement of existing double binds.

One of Bateson's (1972) major hypotheses is that metaknowledge follows knowledge. The pupils do not only construct knowledge, they also construct metaknowledge. The didactical implication of this is that the relationship between learning and metalearning should be discussed again and again with the pupils.

A double bind occurs only when there is no metaknowledge about the contradictions in the messages passed on to the pupil. A teacher can be aware of such contradictions without mentioning them to her pupils. One reason for this is that she believes this might weaken her own position. A familiar example of this is when the teacher has insight into the destructive effects of a grade system, at the same time as she tells her pupils to play up to it.

In this paper I have focused on the importance of keeping in mind the goal of a learning activity. Double binds may occur if the pupil is expected to learn without having in mind why she is learning the particular field of knowledge. The idea of repeatedly reviewing the history of a project with the pupils will enforce reflection on the goal of the project. In its turn this increases the pupil's possibilities of making herself independent of her teacher, as s/he becomes aware of the present context of independence.

If a field of knowledge is passed on to the pupils during a series of separate lessons, the familiar method of opening and closing each lesson with reviews of what is going on in the context of purpose, can combat the production of double binds. Such reviews might be of value for the teacher as well because they may help her to keep in mind the relationships between the different parts of her didactical assignment.

NOTES

[1] I use "individual" and not "individuals". The natural thing to do is to use plural as the teacher usually addresses groups of pupils rather than one particular pupil. If I do this, however, I shall have to introduce the dialectic between the individual and the group. So far, this is premature, at least for this writer. A discourse where the subject is the learning group rather than the learning individual will in this particular context analyse double binds of schoolclasses rather than double binds of individuals.

[2] The validity of such a text as research data is based on an assumption that the text reflects the cognitive map of its author. Its value for analysis is furthermore dependent on whether this cognitive map is the individual's guide for practice. When a text describes a problem solving activity its validity thus rests on an assumption that it reflects the activities it describes (Bråthen, 1982: Hammersley and Atkinson, 1986).

In the present case the texts were delivered as examination papers. It cannot be disregarded that the authors to a certain degree played up to what they believed to be the expectations of their tutor. In this case the texts nevertheless have some significance as research material, as they give evidence about the problem solving strategies of the authors.

Such texts can be of value even if they do not report what has happened in the classroom. As texts they are representations of the teachers' language. In this paper the possibilities of a didactical tool, the double bind, is examined. The texts thus provide an opportunity to justify the uses of this tool and the theory which produced this tool.

[3] Again I have to stress the limitations of my discourse. I write here about the independence of an individual in relation to a goal directed activity. A complete didactical analysis would of course extend the discourse to treat the independence of a group in relation to a goal directed activity.

REFERENCES

d'Ambrosio U.: 1985, *Socio-cultural Bases for Mathematics Education*. Unicamp, Campinas, Brazil.

Bateson, G.: 1973, *Steps to an Ecology of Mind*, Paladin Books, Herts.

Bruner, J.: 1983, *Child's Talk*, Oxford University Press, Oxford.

Bruner, J.: 1985, 'Vygotsky. A Historical and Conceptual Perspective', in J. Wertsch (ed.), *Culture, Communication and Cognition, Vygotskian Perspectives*, Cambridge University Press, London.

Bråthen S.: 1982, Kognitiv kartlegging og gyldiggjøring. I Holter og Kalleberg

Kvalitative metoder i samfunnsforskning, Universitetsforlaget, Oslo.
Gerdes, P.: 1986, 'Conditions and Strategies for Emancipatory Mathematics Education in Underdeveloped Countries', *For the Learning of Mathematics* **5**, 15–20.
Greenfield, P.: 1984, 'A Theory of the Teacher in the Learning Activities of Everyday Life', in B. Rogoff and J. Lave (ed.), *Everyday Cognition: Its Development in Social Context*, Harward University Press, Cambridge Mass.
Hammersley M. and Atkinson P.: 1986, *Ethnography, Principles in Practice*, Tavistock Publ., London.
Howson, A.G. and Mellin-Olsen S.: 1984, 'Social Norms and External Evaluation', in B. Christiansen, A.G. Howson and M. Otte (ed.), *Perspectives on Mathematics Education*, D. Reidel Publ. Co., Dordrecht.
Johnsen Høines M.: 1987., *Begynneropplæringen*, Caspar, Nordås.
Lindén, N.: 1989, *Stillaser om barns læring*, Caspar, Nordås.
Mellin-Olsen S.: 1987, *The Politics of Mathematics Education*. D. Reidel Publ. Co., Dordrecht.
Mellin-Olsen S.: 1987a, *The Role of Thinking Tools in Mathematics Education*, CIEAEM, University of Southampton, Southampton.
Mellin-Olsen S.: 1988. Om å sette kunnskaper fri. I Mellin-Olsen (red.) *Omkring kunnskap*, Bergen Lærerhøgskole, Bergen.
Mellin-Olsen S.: 1989, *Kunnskapsformidling*, Caspar Forlag, Nordås.
Sinclair H.: 1983, 'Young Childrens Acquisition of Language and Understanding Mathematics', in *Proceedings of the Fourth International Congress on Mathematical Education*, ICME, Birkhauser.
Vygotsky L.S.: 1978, *Mind in Society*, Harward University Press, Cambridge, Mass.
Wood D., Bruner J., and Ross G.: 1976. 'The Role of Tutoring in Problemsolving; *Journal of Child Psychology* No. 17.

ENTR'ACTE 3

In the next chapter, Dörfler moves the reader to consider further the relationship between the learner and mathematics knowledge. There is broad agreement that the essential characteristics of mathematical knowledge are its generality and abstractness, and accordingly, developing the processes of abstracting and generalizing is a prominent goal of teaching and learning mathematics. Despite this central role, the meanings of the concepts "abstract", "general", "abstraction" and "generalization" remain all too often vague and implicit, and it appears that an adequate didactical-psychological theory is missing, which might be helpful in very many didactical situations. Therefore a theoretical model for the processes of abstraction and generalization is at the heart of this chapter's contribution. The essential features of this model are the emphasis on actions as the genetic source for abstractions and generalizations, and the description of cognitive tools which enable the learning subject to construct abstractions and generalizations. The broad applicability of the model is demonstrated by examples from school mathematics and advanced mathematics as well. Further, the model is contrasted with different conceptualizations of abstraction and generalization which appear to be pervasive in the teaching of mathematics in school. The inadequacy of those points of view for capturing the specificity of mathematical concepts is discussed.

W. DÖRFLER

FORMS AND MEANS OF GENERALIZATION IN MATHEMATICS

Be wise generalize!

1. INTRODUCTION

Both in everyday and in scientific thinking generalizations have great importance as (general) concepts, statements, declarations, ideas, hypotheses, argumentations etc. As a matter of fact the importance of generalizations is not only restricted to individual thinking but refers equally to social communication. Generalizations are both object and means of thinking and communicating. Generalizing and the resulting generalizations in colloquial language differ, as we know, from generalizing in many sciences, but generalizing in mathematics certainly takes specific forms and makes use of very special means. These forms and varieties as well as the means of generalizing are the objects of the subsequent reflections.

Although one can probably not define such basic concepts as "generalizing" or "generalization" without using tautologies or "circularities", I want to, and possibly have to, try to give an explanation of what I mean by these terms. I understand generalizing as a social-cognitive process which leads to something general (or more general) and whose product consequently refers to an actual or potential manifold (collection, set, variety) in a certain way. The kind of reference to a closed or open manifold and the way of its mediation depend on the particular form of generalization.

Generalizing can be seen as a psychological process within the cognition of the individual, whose products are then the corresponding cognitive constructs (schemes, frames). But this individual process is always conditioned and mediated socially, as it uses and depends on means attained and prepared by society, like language. One can hardly say anything about these psychological processes directly except from what can be observed in social communication (for example in interviews); they themselves are not observable. Compare Davis (1984) and Phillips (1983) for these problems. Nevertheless there is a point in setting up scientific theories for these processes. Apart from the subjective-psychological level there is on the other hand the objective-epistemological level of generalizing. This is the level of the objective, socially-shared knowledge, which is the common basis for the successful communication by individuals. This is the more-or-less standar-

Alan J. Bishop et al. (eds.), Mathematical Knowledge: Its Growth Through Teaching, 63–85.

dized presentation of knowledge which you find in books, lectures, discussions and so on or what you can also state analytically. An analysis at the epistemological level certainly gives no information about the inner structure of psychological processes but it does make possible statements about their outer functionality and their effectiveness concerning the observable behaviour.

My studies refer to this epistemological level because I am analyzing mathematical knowledge with regard to generalizations and the processes which lead to it. But I should like to stress that an epistemological analysis can and will give hints as to what is also psychologically relevant. Its results can be used as the basis for developing models for cognitive processes leading to the type of knowledge analyzed epistemologically which then can be evaluated empirically. Along such epistemologically-based psychological theories didactical proposals can be developed which again through their practical implementations in the class-room offer opportunities for empirical evaluation. In this sense on some occasions in this paper I will dare to draw psychological consequences too.

Before turning to the main part of the paper, I want to give some of my reasons for treating this topic as an object of research.

a) As a result of the project "Development of Formal Qualifications in Teaching Mathematics" we found generalizing combined with abstracting as a central ability and a common basis in a lot of mathematical activities. According to the kind of activity, generalizing is found in a specific form. Compare Dörfler (1988, 1984a, b) and Peschek (1985, 1986) for the project mentioned.

b) In didactic discussions the term "generalizing" or similar terms are used very often. But this is mostly done in some naive way by relying on the intuitive understanding of "generalizing" and by assuming common sense concerning the meaning of this word. In doing so one ignores for example that there are also fundamentally different forms of the process as well as of the products of generalizing. Many proposals for teaching are actually proposals for developing processes of generalization at the level of content. But these processes themselves are not analyzed any further. In my opinion we need a didactics or a didactical theory of generalizing.

c) I think that we will learn to understand the genesis and the character of mathematical concepts better if we analyze the underlying processes of generalization and abstraction more exactly. Generality is attached to all the mathematical concepts, objects, propositions, proofs etc. on a large scale. In other words I want to understand mathematics better in its quality as a product of socially determined, generalizing human thinking.

2. GENERALIZATION BY COMPARISON

The first form of generalizing which is going to be treated is not only the traditional and classical perspective of generalizing but also the form which is, at least implicitly, the basis of many didactical proposals for teaching. Nevertheless it is in its most extreme form a kind of antithesis to the forms of generalizing which are going to be dealt with later and which are considered to be constitutive for mathematical concepts.

This form refers to the process of developing generic terms which was already described by Aristotle. The basic process is to find a common quality or property among several or many objects or situations (mostly from sense perceptions) and to notice and record these qualities as being common and general to these objects or situations. The common is found by comparing the objects and situations with regard to their outward appearance and their figurative shape. Many concepts (i.e. generalizations) in the empirical-descriptive sciences could have been developed in this way. Therefore, this kind of generalization is also called empirical generalization. The common quality (or system of such qualities) is isolated mentally and is detached from the objects or situations. This process is called empirical or Aristotelian abstraction.

In mathematics this would mean for instance that the concept "triangle" represents the common and general property of triangular objects. The commonality is developed psychologically by the individual by comparing many concrete triangular shapes with non-triangular shapes. Thereby triangularity (having 3 vertices) is realized as the only distinctive feature common to all the objects in a subclass of the scanned objects. This common property is then detached from the objects (abstraction from the concrete instantiations of triangularity and their many other properties) and constituted (mentally and epistemologically) as a general property (a concept) which can be possessed or not by objects. It was for many years (and is?) a permanent point for discussion in philosophy in which sense this general property "exists" independent of triangular objects (problem of universals). Can we think of triangularity without in some way thinking of or imagining concrete triangles? From a psychological point of view this kind of triangularity can be considered as the mentally stored and reflected process of perceiving triangular objects with specific attention focused on the "being triangular".

In the teaching of mathematics at all levels the empirical approach to generalization and abstraction is very often the foundation for the teaching methodology used. The teacher demonstrates the intended concept as the specific property of several so called introductory examples. Take for example the concept of a function, of convergence, of the derivative or even that of a fraction. In many school books you will find wordings like: "All our examples show this or that property and if this property holds we will call…"

Such a perspective of seeing the general as the common soon becomes problematic with other mathematical concepts. For which objects does an ellipse represent a common quality? How can one explain "ellipse" as developed by observing and comparing? What and how should one observe and compare? We can only perceive with our senses that a form is similar to an ellipse, but there is no means of perceiving the mathematical specifics of an "ellipse". However, most of the didactical proposals for achieving concepts as empirical generalizations are not that naive. But often teaching tends to offer as many situations as possible – as examples for the generality that the pupil has to acquire. Out of these situations the common and general as the content of a concept should be developed. In this way one postulates implicitly that generalizing can be mediated by comparison and empirical abstraction. In doing so, an analysis of the particular situations is not in the foreground, but rather a comparison of them is.

Comparing and looking for common features among different objects or situations is clearly a kind of activity but this activity is in no way related to the common and general structure or property one is looking for. The comparison will mainly serve as a means of selecting those features which are common to (many) objects for the purpose of abstraction and ensuing generalization (i.e. the constitution of the common as a general and as an entity of thought in itself). But in principle the comparison will be neutral with respect to the concept to be constructed. In this sense empirical abstraction and generalization can also apply to processes which are then compared in the search for common structural features. In many didactical situations this will be the precise case. It is typical of this type of concept formation by empirical abstraction and generalization that the various specific instances of the processes are given, mostly by the teacher, and the students are requested to look for common features. Take as an example the derivative, which is introduced via "examples" like velocity, gradient, density and others. The derivative is then constituted as the common structure of these various processes. But this structure is not developed by the students themselves. I suspect that in many cases the superficial structure of the pertinent formulas is the basis for the comparison of the examples. Rather than the process itself it is the comparison of various instantiations ("concrete examples") which is the source for concept formation. This is in strong contrast to what will be described later as theoretical abstraction and generalization.

It is small wonder that this theory of generalization has been criticized in many ways or completed by other theories. Compare Cassirer (1910), Dawydow (1977), Damerow (1982), Rubinstein (1973) or Beth and Piaget (1966). I cannot deal with these partly philosophical discussions in great detail here, but I do want to mention some short but essential points of critique of empirical generalization:

1. The generality must be considered as being already present in the objects and situations; then it is "only" abstracted and thus detached from them. The general is detected and not constructed, and it must in principle be given to the senses. But the question remains: Where do the qualities which are stated as "common" actually come from? How can one imagine a comparison e.g. without the learner's already existing knowledge of the relevant qualities? What should be developed in the process of generalizing thus seems to be presupposed. One can imagine such a process of developing a general quality at best with eyecatching qualities: Rubinstein talks about "strong" qualities which stimulate the development of everyday concepts in colloquial language like shapes, colors, types of plants and animals and many others. On the other hand one will have to presume that in many other cases the generality is stressed, pointed out (thus abstracted) and given by a kind of teacher. The generality may also be noticed in one of the situations, and then searched for in others deliberately. Therefore, I believe that we need such a stereotype or a template which helps us to compare objects and situations. Conclusion: Rather than the objects and situations themselves being compared, they are contrasted by means of a stereotype or a template which has to be developed in advance and which develops further gradually. But the "fitting" of such a template and the determination of it poses a crucial problem which will be dealt with by my other forms of generalizing (adjustment and fitting will be interpreted as the executability or feasibility of certain material or mental actions).
2. The generality thus formed depends totally on the field of the objects out of which it was developed. The meaning of such a generality is purely extensional, and consists only of objects and situations which have the quality constituted as the generality.
3. The generality which is formed by comparing in this way cannot be developed or even generalized further, since it is principally restricted to the field of the objects which it comes from and which represents its significance and meaning. Such a generality is limited to those objects whose common quality it represents. A triangle or a circle as an empirical-geometric form, cannot be generalized further.
4. Different objects and situations frequently have many qualities in common. How can one decide which common qualities are the intended ones? So-called counter-examples can help here only in the viewpoint of the one who already knows this generality and can "see" it in the objects: After all, they can correspond with the positive examples in other respects. Empirical generalization by looking for common properties lacks a specific goal or intention to decide what is essential. The generality thus formed is somewhat arbitrary: Why is this and not another quality seen as common? The common, and the eyecatching common too, can be irrelevant for the objects to be compared. Why

should a general quality be formed at all as a common quality? The particular situations could already be interesting enough, especially for children. Studies by Luria (1976) show very forcefully that our conceptions of the general depend on development and evaluation and that they must be seen as relative to the context of people's activities.

5. Pure comparison does not give the reason as to why the common is common. Further, in many cases this common cannot be stated explicitly or described, except by giving a (paradigmatic) example. This means that only the extension but not the intension of the corresponding concept is determined. Therefore the empirical generality entails no rules for the establishment, whether it refers to an object as a quality or not. This can often only be decided by comparing the object with other objects which have already been subsumed under the general (or with their imaginative pictures) and which can serve as a kind of template or representative of the generality.
6. In the didactical process one cannot really examine whether the learner has mentally formed or not the generality which is intended. Because of the way of developing the generality an examination is only possible by showing so-called examples and counter-examples. But "correct" decisions can also be achieved through "wrong" means (another generality). Another source of error is that the learner puts qualities into the generality, which are more or less incidental to the underlying field of objects (besides the general intended). Such qualities are always present! For example, geometric forms which are formed via special positions of the figures: A square is not recognized as a square if it is depicted as "standing" on one of its vertices.
7. This empirical theory of generalization does not tell us what the learner should do in order to construct the generality cognitively. To some extent, the ability to recognize the generality is postulated. Therefore such a theory cannot produce any didactical hints about how to organize instruction, except to make comparing possible through as many different situations as possible. Even some formalizations of generalizing stick to this framework, see Bauer (1978).

Being confronted with these and other critiques and deficiencies of empirical generalization as a model for the development, and as genetic source, of general concepts one has to ask the question: how could such a point of view exert so strong an influence in philosophy and also in didactics. Though this model is related to other philosophical, epistemological and also psychological positions and theories there might be two main reasons for its "success". If one starts from some given general property or concept it is always possible to interpret it as the common (and possibly essential) feature of a certain class of phenomena (objects, processes, situations). Thus in hindsight, any generality can be viewed as a common aspect which can then be further viewed as to be abstracted (i.e. isolated, accentuated, detached) from other features. Yet this only exhibits a necessary quality of the generality but does not tell us in many cases anything about the genetic process that lead to the generality. I think the main flaw in this type of argument is that it is used by those who already know the generality and its instances and who have already abstracted and constituted it as a kind of mental entity.

Another reason for the persistence of the empirical model of generalization is the fact that empirical generalization and abstraction are the source for many general conceptions in practical life and even in empirical science. It is the method and means for classification, for organizing and structuring everyday life experiences. In this sense the critique cited above does not deny the importance of processes of empirical generalization which often lead to important conceptions which then guide our perception and even many practical activities. For instance in some situations it is important to distinguish a poisonous mushroom from an edible one or a fork from a spoon just based on their shape or other empirical property.

It should be further remarked that the processes which I will characterize below as theoretical generalization need a starting point consisting of objects and their properties which in many cases will be the result of empirical generalizations. On the other hand often a theoretical concept will "explain" why an empirical concept is useful and meaningful or why the respective generality is common to many different objects or situations. Thus there is a mutual relationship between the different forms of generalization and they are separated here mainly for the purpose of a clear analysis. Another reason for this separation is that the mathematical conceptions which are studied by mathematical theorists should be viewed as having their genetic source in what is termed here "theoretical generalization" because otherwise their epistemological status and properties are difficult to understand and explain.

3. GENERALIZING SCHEMATA AND INVARIANTS OF ACTIONS

In this section a model of generalizing is going to be dealt with which demonstrates the essential features of those processes which in my view can lead in many cases to genuinely mathematical concepts. The later Sections 4 and 5 will show that by using the model developed here one can interpret different, objectively determinable processes of generalization in mathematics.

The processes of generalization described in the Sections 3 to 5 share a common overall schema which of course might vary slightly depending on the particular concrete cases. The main differences between the various processes result from which particular elements, or aspects of an action, or a system of actions, attention or interest is focused upon. This focus of interest might be the schema (process form) of the actions itself (Section 3), the product or outcome of the actions (Section 5), or conditions for the feasibility of the actions (Section 4), formulated as relations between or properties of the objects. Depending on this focus of interest different relations will be constructively abstracted by symbolizing the objects which are put into the respective relations by the actions.

Following the tradition established by Rubinstein and Dawydow I call this form of generalization 'theoretical generalization' though the term 'operative generalization' would actually be more appropriate.

Another remark is appropriate here. The informed reader will readily realize that the model of theoretical generalization bears a rather strong resemblance to certain Piagetian ideas. A common starting point surely is to consider actions as the genetic source for mathematical conceptualizations. Piaget uses the term "reflective abstraction" for such processes but does not pay much attention to the cognitive tools by which reflective abstraction is, or might be, accomplished. This is one respect in which the theory developed here goes beyond the traditional Piagetian approach. Use of symbols and the introduction of variables are exhibited as important means for abstraction and generalization. Further, Piaget seems to be interested mostly in the explanation of how very fundamental mathematical (or rather logical) concepts derive from actions via internalization in the form of operations which additionally have to be organized in a certain way (groupings of operations).

My theoretical approach intends to be in principle applicable to all kinds of mathematical concepts at all levels of abstractness and generality. This necessitates a broad concept of actions which, for instance, includes symbolic actions as well. Another point is that for Piaget the interiorized actions, i.e. the operations, reflect mainly the schema of the actions. To attain our goal we have to extend the approach to what are called here the conditions of actions and their results. In other words, our theory and the various examples show that mathematical concepts can be reconstructed genetically from actions in several distinct ways, only some of which are covered by the original Piagetian approach. Thus the model developed here is more specific (by modelling in detail the process of what Piaget called reflective abstraction) and broader as well (by including a great variety of mathematical concepts).

As another remark I want to state that my model for generalizing is to be seen as reconstructive: the generalizations (concepts) as given products can be conceived as being obtained and constructed through processes as they are described here; and that holds both psychologically (by the subject learning) and epistemologically. However, I do not exclude the idea that the same product can be obtained in another way too. But I do think that the schema representing a concept cognitively comprises not only the static, completed product of processes of generalization, but also its process of construction (in condensed form). This means that the subjective meaning of generalizations (concepts) does not (only) consist of their formal structure but of their complete genesis and development.

To begin with I would like to describe the process and its means by which this form of generalizing is characterized. Then I will give some illustrative examples.

The starting point is an action or a **system of actions**, which are material, imagined or symbolic, but always concrete actions. In particular, the elements of these actions are certain (material or ideal) objects. The objective, means and course of these actions direct one's attention to some relations and connections (constituted by those actions) between the elements of the actions. This is especially true of the planning or evaluating reflexion of actions. These relations prove to be steady when the actions are repeated (as often as you like) and I will call them **invariants of actions.** In many cases the actions in a certain and invariant way combine the original elements (which I consider as a special form of transformations). This can be done physically or by establishing certain connecting relationships. Since at least from a certain point of view these combinations and basic transformations are the essential and characteristic features of the action, I will call them the **schema** of the action. The schema is therefore a special invariant and in the following I will not discriminate between the two terms.

Even stating these invariants needs a certain **symbolic description**, because one has to introduce symbols for the elements of the actions or for the quantities relevant to them respectively, and for the transformations or combinations on them induced by the actions. These symbols can be of a verbal, iconic, geometric or algebraic nature. In any case the invariants are described by means of these symbols and are thus fixed. At first these symbols are symbols for the elements of the action, or for their numerical or geometrical characteristics (if these are relevant for the actions), and for their transformations. Therefore the symbols have the quality of representatives first, and play only a descriptive role. They are often substituted by **prototypes** of the original elements of the actions. The prototypes correspond with these elements in the qualities and relations which are relevant for (the executability of) actions, but are otherwise simpler than these. This stating of invariants and their symbolic description have the character of a **process of abstraction** since certain properties and relationships are pointed out and attention is focused upon them. By this means they gain independence to a certain degree from the elements or objects with which they are originally associated. It should be emphasized that this is a **constructive abstraction** in that what is abstracted is constituted by the action and gains meaning and "existence" via the actions.

In many cases the elements of the action, and perhaps the action as well, have a certain variability themselves in the starting situation, because they have to be of a certain "type", which reflects e.g. in the prototypes. The elements of the action and to some extent the actions themselves can be substituted or thought of as substituted by others, whilst the type remains the same. Thereby – and this is important – one gets in each case the same invariants, represented in each case by the respective symbolic description.

From the beginning, therefore, the symbols have a certain range of reference and this range can now be extended gradually. This extension of the range of reference is done by asking: by means of what, and with what, and how, can one act so as to maintain the invariants already stated? The invariants in their symbolic description serve as a kind of search-screen for the possibility of "identical" actions in this sense. This is a matter of replaceability, interchangeability of the elements of the action, and also of the actions themselves, whereby the "structure" of the actions stated through the schema and the invariants should be maintained. This is a kind of **extensional generalization**. First this generalization is achieved through similarities in the objects which can be elements of action (empirical generalization). But later the invariants will become the criteria for choosing objects as elements of actions. Therefore the symbols used in the beginning are **variables with substitution-property.**

By reflecting on the symbolic description of the invariants the symbols used will begin to substitute gradually the elements of the actions and the tansformations with them. The symbols get the character of objects; they become elements of actions themselves and as such representatives or "carriers" of the invariant relations respectively. One can talk about a detachment of the symbols: the symbols are detached from their range of reference in a kind of changing of perspective. They become independent objects which gain their significance and meaning from the relations (the invariants) and from the operations for which they act as symbolic objects. The symbols become objects (objectively and in the cognition of the learner), whose determining qualities are the invariants descriptively stated earlier. In this way a new object of thinking develops, a mathematical object whose intensional significance and meaning lies in the invariants. This mental object is a generality, because it has a potentially unlimited range of reference. In this object the symbols are now (primarily) **variables with the character of objects**. This reification (objectification) of the variables and of the symbols completes the process of abstraction which began with fixing the invariants. In this way the invariants are detached from their original "carriers"; they gain independence and form the abstracted generality. Here abstraction is the means for developing an **intensional generalization.**

The constructed generality, expressed by the use of concrete object-like variables (for elements of actions and for actions), subsequently serves, in a subjective-cognitive way as well as in an objective-epistemological way, for further **extensional generalization**. By appropriately interpreting the variables, new systems of actions are subsumed under the generality whose range of reference is thus extended further. This may result in constructing new systems of actions by means of the generality such that the invariants expressed in the generality are present and "valid" there.

Although I know about the dangers of such descriptions because of the simplifications connected with them, I would still like to offer a graphical schema (see p. 74) for the process described previously.

Before referring to examples, some important features of this process of generalization should be pointed out which apply to the processes considered in Sections 4 and 5 as well:

- Generalizing in this form basically needs only one starting situation or one type of starting situation, from which the generality can be constructed through abstraction of the essential invariants in the context of the system of actions.
- The generality contains rules for its use; it has both descriptive and operative character. The operative character results from the genesis out of the actions.
- Generalization takes place with respect to a system of actions: The elements of the actions, and the transformations exerted on them, are conceived as variables while a certain system of relations is maintained (the invariants).
- The system of actions determines to some extent the direction and the content of the generalization (the invariants), whose relevance is based on the individual's activity.
- The generality thereby obtained does not represent the qualities of things (which have been stated), but relations between things which have been established, constructed by actions and which only become meaningful, and in existence, through them.
- Here the generality is fixed intensionally or intrinsically and is not limited to this "example". At the same time it is an instrument for extending its range of reference by guiding and directing human activity.
- Due to the operative character of the intended generality one can conceive of tasks which permit the acquisition of some evidence about the degree to which the student has succeeded in constructing the general concept. This is a marked difference to the generalization by comparison.

One can rightfully ask the question, where do the actions or the starting situations come from? Usually there will be a kind of problem which gives rise to the actions, or the system of actions has already been carried out by somebody and somebody else starts to analyze it. In this sense it is not necessarily one's own actions, but actions can be observed which are mentally reconstructed and reflected upon. For a didactical implementation of course the problem of motivation crops up and this is indeed very crucial. How can one get students to carry out (or imagine) actions and to investigate them? I can't offer an answer and possibly it is the teacher who must select appropriate starting situations since it is only she who knows the students and their interests. It could be the task of didactics to develop as many as possible different examples for starting situations. Here I want to stress that

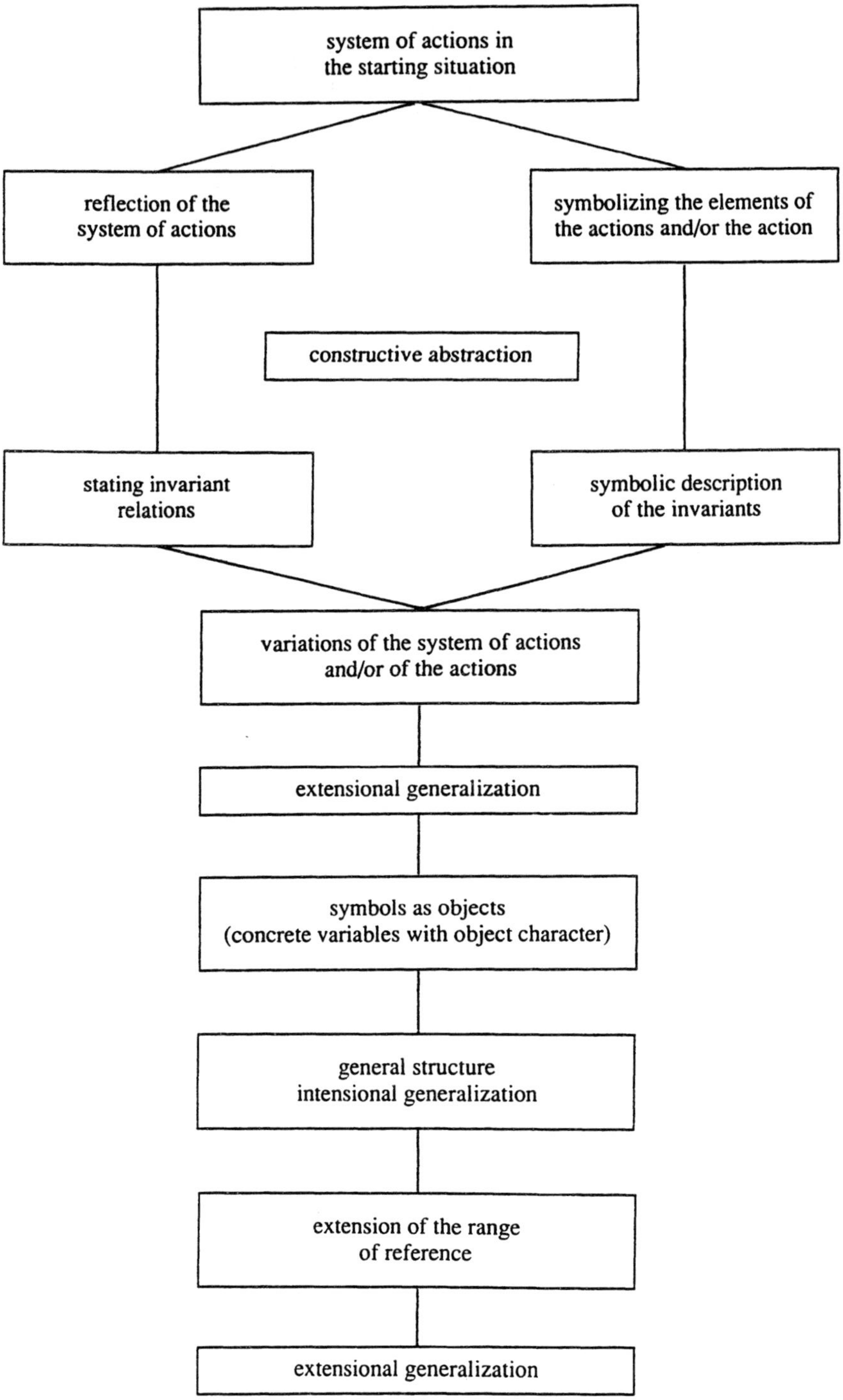
system of actions in
the starting situation
reflection of the
system of actions
symbolizing the elements of
the actions and/or the action
constructive abstraction
stating invariant
relations
symbolic description
of the invariants
variations of the system of actions
and/or of the actions
extensional generalization
symbols as objects
(concrete variables with object character)
general structure
intensional generalization
extension of the range
of reference
extensional generalization

the following examples like those in Sections 4 and 5 are not meant to be transferable to the classroom in any straight-forward way. For this a detailed didactical transformation would be necessary.

Example: Algebraic Terms. According to the schema developed here, one has to choose a starting situation which essentially consists of a system of actions. Assume that one carries out a certain arithmetic (or algebraic) calculation in an appropriate number range, like say (2*3+5)/((2/5)*7-6). With regard to the process of generalization of interest here it is of minor relevance where the actions come form, and in the other examples too we will start from an already executable (at least mentally) system of actions. After having carried out the action (i.e. calculation) possibly with varying specific numbers one will try to express the schema of this action. This can be done verbally by describing in natural language what one has done: First we multiply such and such, to the result we add ... To enhance the shift of attention to the schema of the action symbolizing by letters of the operands (i.e. elements of action) will be advantageous. These symbols will at first only refer to a certain range of numbers. Other symbolic representations of the action schema will support the process of abstracting the schema from the specific actions: tree representations of the calculation, flow charts and the like. The use of calculators might support this process since it allows the learners to focus attention on the operations themselves and their composition in the actions. In parallel a further extensional generalization can take place by extending the calculation, for instance to arbitrary real numbers. The decisive step is then to view the symbols introduced themselves as the operands of the calculations for which one stipulates the executability of the actions as the defining property. This will establish the schema of the actions as a general structure (intensional generalization), namely as the term which is open to all interpretations which fit into this structure. The ensuing extensional generalization of this term can consist of admitting very different mathematical objects as substitutes for the objectified symbols: complex numbers, (regular) matrices, (invertible) functions and many others.

In this example the process of generalization was confined to the schema of the actions: determining invariants often necessitates reflection on a more complex system of actions which relates certain subsystems, for instance with regard to their effects being equal. This kind of equating of different terms establishes algebraic formulae like the law of distributivity. As a result of an appropriate process of generalization these laws will not be restricted to qualities or rules for calculations in certain number fields but will constitute fundamental structural relations. Thus $a*(b+c) = a*b + a*c$ is "the" law of distributivity with a, b, c being uninterpreted object-like symbols and $*$, $+$ being uninterpreted calculation signs.

Example: Geometric Construction. Consider a rectangle and a certain way of constructing it. The latter can then be viewed as a system of actions and can be taken as a starting situation according to our model of generalization. The elements of the action here are the points, lines and angles (and possibly others) drawn and/or imagined, and they are to be combined in a certain way to obtain a rectangle. An initial range of extensional generalization is therefore quite natural. The schema of the construction can be stated verbally or by introducing symbolic means. For instance, an appropriate LOGO program could serve as a symbolization of a certain way of construction. Applying the process of generalization to any of these symbolizations will lead to an abstract and general formal structure in which certain elements of the original construction gain the character of objectified variables (either only points; or points and lines; or points, lines and angles). This then permits us to interpret these variables in very different settings thereby defining what in the setting by a rectangle should be understood. How farreaching this process of objectifying variables is, will depend on one's interests and motivation. In any case it will support the shift from viewing a rectangle as an empirical figure to viewing it as a theoretical construct based on the (idealized) way of how to obtain the figure.

Example: Groups. There are of course many different systems of actions as starting situations in our sense for this concept: it is only natural that they are didactically suited differently. The rotations of a square (mod 2π) may serve as an example for our general position. These rotations are the elements of the actions and the actions in the system are the compositions of the rotations. Relations between the rotations can be stated as invariants, such as (in algebraic language) associativity, neutral element ("neutral/zero action") and inverse element, but also (here!) commutativity or cancellation. These invariants are stated in the course of the reflection on the whole system of actions and they are noted symbolically by means of symbols for the elements of the actions, i.e., rotations, and the actions, i.e. their compositions. The phase of initial extensional generalization here could be the transition to other regular n-gons with a fixed or even an arbitrary n, or the extension of the system by including reflections. Objectifying the symbols as concrete variables turns the generality thus constructed into the formal structure of a group. Other starting situations: permutations with composition, number domains, residue classes, matrices etc. These situations can also serve as examples for extending the range of reference of the constructed generality such that all phases of the model can be traced here. These examples show very clearly how the invariants stated in the starting situation become "arithmetical rules" or (more general) rules for acting when dealing with concrete symbols or variables respectively.

Example: Word Problem. Here it is a matter of constructing what one can call the general type of a word problem or its abstract arithmetical structure as a generalization by means of a process corresponding to the model described above. In this connection only word problems which have an operative solution through (possibly imagined) actions are tractable by our model. Consider the following "problem":

A man and a woman walk through the desert. If their rations of water per day remain steady they will manage with their water supply eighteen days. The man alone would manage twenty-four days. What proportion of the daily ration of the man does the woman drink?

We now suppose that a solution via (imagined) actions has been devised which will then serve as the starting situation for the process of generalization. Starting from the given facts (both together drink the same in 18 days as the man in 24 days) one possible solution (by actions) is to pour the additional 6 rations of the man (RM) into the 18 rations of the woman (RW), so that the different rations become elements of the action. Thereby the following relation (invariant) can be stated (perhaps in addition to others):

$$6 \text{ RM} = 18 \text{ RW}$$
$$1 \text{ RM} = 3\text{RW}.$$

Besides this algebraic symbolization there are of course verbal and iconic descriptions. A verbal description could be: The man drinks in 6 days what the woman drinks in 18 days. Or: In the number (6) of days which the man alone can survive longer than both together, he would drink what the woman drinks during the time (18 days) they are together. An iconic description could be:

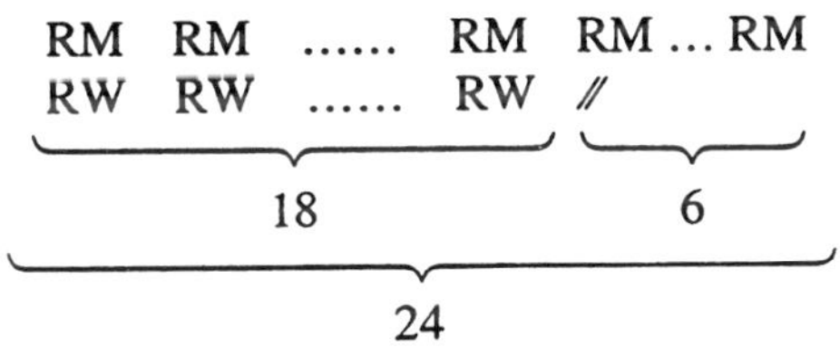

or more geometrically:

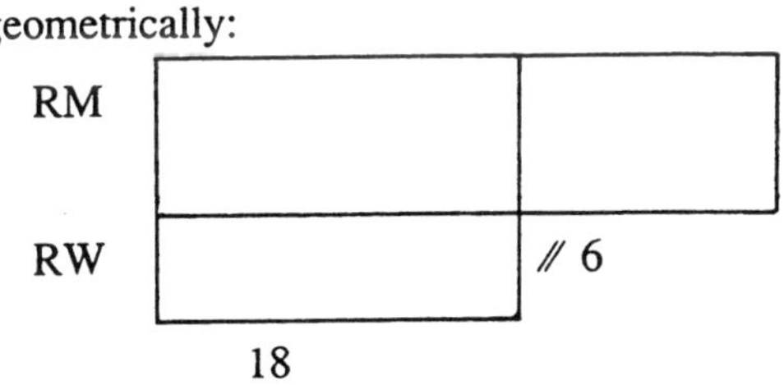

Another possible relation follows from equating what they drink together in 18 days with what the man drinks in 24 days:

$$24 \text{ RM} = 18 \text{ RW} + 18 \text{ RM}.$$

This too can be represented in a similar way by rectangles. One should realize that these relations are consequences of the actions which relate the various magnitudes.

Now the process of generalization with its different sections applies to the invariants thus described. Thereby the numbers 6, 18, 24 and the rations RM, RW can and will be transformed into variables. This can be stated verbally by using phrases like "some number of days". A first phase of extensional generalization could lead to a formulation like this: The total amount of some commodity can be placed into m containers of size A plus m containers of size B or just into n containers of size A. What is the ratio of size A to size B? Or more generally: m copies of A plus m copies of B are "equivalent" to n copies of A. What is the relation of A to B? In this formulation the gradual detachment of the variables A, B (and possibly m, n) from the original field of reference is visible and can be enforced.

The structure (type of the problem) can such be formulated in algebraic description as:

$$mA + mB = nA \text{ or } (n-m)A = mB.$$

The reader will easily supply a geometric representation.

Here I want to stress once more that at this stage A, B are concrete, object-like variables. This means that they do not have a given range of variability over elements by which they can be substituted; but they are independent objects which can be manipulated according to certain arithmetical rules. Their range of reference are all the ranges of those elements, where there are such possibilities for manipulating and operating (e.g. discrete sets, numbers, extensive quantities, vectors, matrices, functions and the like). It is left to the reader to carry out a further extensional generalization by inventing or constructing other problems which carry this general structure.

4. GENERALIZATION OF CONDITIONS FOR ACTIONS

In order to carry out actions, adequate elements of actions have to be there and these must fulfil certain conditions. Quantitative and three-dimensional geometric conditions are especially important for mathematical generalizations. Such conditions can be recorded as qualities of and relations between the elements of the actions. A process of generalization principally similar to that of Section 3 is then applicable to these qualities and relations and will comprise the components:

- recording the conditions of the action
- symbolizing the relations stated (mostly via symbols for the elements)
- variations of the elements of the action (replaceability in regard to the fulfillment of the conditions for the action)
- symbols as concrete (object-like) variables whose determining qualities lie in fulfilling the conditions for the action

- conditions for the action to be the generality, the abstract mathematical object
- extension of the range of reference of the variables and therefore of the generality by constructing and finding fields of elements where the conditions of the action are fulfilled.

The essential difference between the processes described in this chapter and those in the preceding one is only a difference in the correspondence of the relations to the action.

Example: Circle. Besides other possibilities, rotating or rolling a disk around an axis may be seen as the action in a starting situation (while the axis should always remain at the same height above the ground surface).

Here the relevant elements of the action are the following ones: the periphery of the disk and the axis. An analysis of the action shows that the constancy of the distance of the periphery to the axis is the condition for the action. A process of generalization as mentioned above leads to a general structure, which also includes the sphere in R^3, the sphere in metric spaces and so on: all the elements with a constant distance from a fixed element. Here "distance" has also become a concrete object-like variable (in the same way as the composition of rotations earlier), for which certain "rules" to be obeyed can be formulated. As an example for the possibility of very different starting situations, it should be stressed that formally the same system of relations is formed via constructions for the circle (a pair of compasses, thread). Then the determining system of relations must be seen as the invariant for the respective actions.

Similar to the circle many geometric objects can be seen as systems of relations, which are developed through generalization of conditions for actions (or of invariants of constructive actions). The examples given by Bender and Schreiber (1985) may be subsumed here for the most part.

Many other rather simple mathematical concepts can be conceived as resulting from conditions for the executability of certain actions or mathematical operations. The reader will easily continue the following short sample of such concepts:

Composite whole number: you can form a nontrivial rectangle with a respective number of square plates; you can get the number as a nontrivial product. Similarly: even number, square number.

Connectivity of a graph expresses the possibility of walking or driving from every vertex (city, station and the like) to any other. *Planarity* is the possibility of drawing the graph in the plane without intersecting edges.

The stepwise *extension of the number system* (natural-whole-rational-real-complex-and further in several directions) can be organized by means of the stipulation of the unlimited executability of certain operations (subtraction, division, taking limits, solving arbitrary algebraic equations).

These only sketched examples show once more that one has to admit as actions in our theory material actions and mathematical operations as well. Concepts as results of previous generalizations can play the role of elements of actions/operations on a new and higher level. This is also demonstrated by the following examples.

Example: Group. The essential group axioms can be seen as conditions for the actions of solving equations of the form a+x = b or $a \cdot x = b$. If one wants to solve an equation of the form $ax + b = c$, a reconstruction of the conditions of the action by such a process of generalization leads to the most important field axioms as the abstract generality of these conditions of the action.

Example: Concepts Deduced from Proofs. Pickert (1984) shows how concepts can be developed by analyzing proofs. From the proof of the intermediate value theorem, continuity can be derived as a condition for the proof, the concept of field through an analysis of the proofs in the theory of Galois; the concept of the ring through analyzing the proofs of propositions on whole numbers; the same is true for Euclidian rings, principal ideal domains and others; the metric space (Fréchet) derives from proofs of limit theorems.

Now proofs are mental actions with mathematical objects and the theory developed here can be transferred to all the cases mentioned without any problem. A very impressive example of this kind is the concept of the matroid which can be reconstructed through an analysis of propositions about linear (in)dependence as conditions for their proofs. The accompanying general structure has a vast and manifold range of reference: graphs, algebraic dependence, transversal systems, incidence geometries and many others. Compare Welsh (1976) and Dörfler (1979).

Example: Generalizing by Reducing a System of Axioms. This type of generalizing is very specific for mathematics and starts from a structure which has already been recorded axiomatically. The more general structure is formed by omitting one or more axioms. Thus one gets

- from the group to the semigroup
- from the field to the division ring or the ring
- from the square to the rectangle or to the quadrilateral
- from the Euclidean to the Absolute Geometry.

Such generalizations were and are really made and they seem to have a high degree of arbitrariness: Generalization for the sake of generalization. Such a procedure can make itself independent as a method, which is used as a technique for deriving concepts. But it can also be stated that such generalizations are based on an analysis of a (complex) system of actions in regard to the conditions for the actions: Which axioms are necessary for being able to preserve a certain part of the theory, that is to be able to perform certain proofs, constructions and operations? Thereby this type of generalization can be subsumed under the form discussed here, but with the restrictions that the symbolizing and the objectifying of the variables have already been done and that generalization starts from a given generality. However, the important fundamental principle of a relation to (mental) actions remains fully valid. At least the generalization gains sense and orientation from such a relation. It may be interesting that here the resulting generalization restricts the possibilities or the range of actions.

5. GENERALIZING RESULTS OF ACTION

An action or a system of actions respectively always has a result, a product, which can be a material object with certain qualities or an immaterial relation. Objects are products of actions because of their qualities, and in those cases relevant for mathematics these qualities are (approximate) realizations of systems of relations. Therefore let us suppose as a rule that a product of an action is to be viewed as a system of relations, which then exists between material objects, between quantities associated with them or between mathematical objects, depending on what one considers to be the elements of the actions. For instance, if the system of actions is the construction of a geometric figure then this figure is the product of the actions and is characterized as such by the approximate fulfillment of certain relations (which mathematically define the corresponding geometric concept).

A process of generalization which basically follows the model stated in Section 3 can start from this system of relations which is abstracted through a reflection of the action and its results. In this context prototypes for the elements of the action will play a special role again, because they direct the attention to the particular system of relations. The result is a generality which can be understood as a model for the product of the action.

Example: Fraction. As the starting situation we consider the following question. Given two values, say a and b, of a certain magnitude (length, volume, weight), try to exhaust b by a or an appropriate part of it. In other words, one wants to measure b by choosing a as the unit of the respective magnitude. Presuming commensurability, this amounts to finding natural numbers m and n such that $m \bullet a = n \bullet b$ or $b = m \bullet ((1/n) \bullet a)$, i.e. b equals the "sum" of m copies of the n-th part of a. A suitable action might be forming multiples $m \bullet a$ and $n \bullet b$ and to look for values of m and n satisfying the above relation. In any case the result of appropriate actions is a pair (m, n) of natural numbers (or even the symbolic equation $m \bullet a = n \bullet b$) which precisely describes (or denotes) the searched for relation between a and b. To these actions and their result the other steps of the process of generalization are now applicable, which leads to the general concept of fraction (or at least to an important aspect of it).

This example shows quite clearly an important quality of the symbolic notations for the relations established in the course of the action: these relations are not an intrinsic property of the symbols (like m, n, a, b) but they must always be added mentally in our thought. As already emphasized earlier, the meaning of the symbolic denotation resides in the respective actions and the relations induced by them. For the teaching and learning this means that the learner has to experience the symbols as a means for expressing the relational outcome of actions he or she has carried out or observed.

Example: Mappings. Functions generated by (e.g. arithmetic or algebraic) terms establish a system of relations which consists of relating the argument to the value of the function and this is in a way the only really relevant result of the system of actions "calculating the values of the function" (and not the value itself!). The same holds true for the case of geometric transformations, where the image point or the image figure are not the result of the actions (the rotation, the translation etc.), but the relation between the point or the figure and their image is. This relation is induced by the particular action. The process of generalization following our model leads directly to the general concept of mappings as special cases of binary relations.

Example: Angle. Apart from other conceivable meanings of the concept of an angle there is the possibility of subsuming it in our theoretical approach. The starting situation here is rotating say a disk around a fixed point. For any half-line from this point one can somehow mark the starting and the final position. The rotation establishes just this relation "starting-final" between the two positions which will be the "same" for every half-line. One can consider "angle" to be the concept derived from this situation by a process of generalization like the one described above. The usual drawings then constitute a symbolic notation for the angle by the use of geometric symbols (like half-lines).

Example: Differential Quotient. With certain (concrete) actions of measuring, a system of relations develops (proportion of the changes of quantities), which first can be generalized to the difference quotient and when adding further actions (refinement of measuring) to the differential quotient. Additionally to the physical measurement of various rates (like velocity, density, acceleration) which are commonly modelled by using mathematical functions, more idealized or imagined actions such as determining the slope of a curve can also serve as instances for the starting situation or for a phase of extensional generalization. The generalization is achieved primarily by introducing concrete variables for which the operation of establishing a kind of distance must be meaningful. The thus developing generality is the schema of the differential quotient as the rate of local change. An analogous process can be developed for the reconstruction of the (Riemann-)integral (measuring of areas as a generic system of actions).

Example: Generalizing via Parameters. Many mathematical objects (ordered pairs, binary relation, R^2, matrix, triangle, graph and so on) and statements contain implicitly or explicitly a certain parameter which is in most cases a natural number. A widespread technique or method of generalizing is to set this parameter variable (over a well-defined or even open range of reference). Similar to the dropping of axioms one can develop a perspective on this method which is oriented along actions. In many cases the particular parameter is an important element in actions which create perceivable objects (mostly symbols). By these objects the particular mathematical object is "realized" as a system of relations: writing two numbers as a pair; writing a matrix (two directions: rows, columns); connecting two vertices in a graph etc. In other words: the value of a parameter is implicitly or explicitly an element in the system of relations, which can be seen as one result of the particular action; the value of the parameter is a constitutive part of the result of the actions. Our process of generalization can now be applied to this system of relations but just with regard to the parameter. It yields a generality which is the schema of such results of actions (and actions themselves) with an arbitrary value of the parameter: n-tuple, n-ary relation, R^n, hypergraph, n-dimensional matrix, n-gon and so forth. The "general action" does belong to this generality as well as an action schema with the particular parameter as variable! At any rate, it seems important to me, that the generalization does not only take place figuratively, but is directed and motivated by actions.

6. SUMMARY

In conclusion I want to sum up the essential constituents, which played both a subjective-psychological and an epistemological role in the process of generalization as discussed in the previous sections. Among them there are:

- *Actions* (material or mental). They direct and control the attention and focus it to certain qualities of, and relations between, the elements of the actions. These qualities and relations can be in different proportion to the actions, as schemata, as invariants and others. The actions can be actions that really took place or actions which are imagined or intended, planned. The concept of the action and the concept of the elements of the action is so comprehensive that even mental and symbolic operations (such as carrying out a proof) fall within these concepts.
- *Symbolizing*: use of letters and artificial signs, iconizing, verbalizing and the like, of the abstracted qualities and relations. In this connection symbols (verbal, iconic or algebraic) for the elements of the action and for

the relations induced (strictly speaking: mentally constructed) between them (through the action) are introduced.

- *Prototypical Elements of the Actions*, which essentially have got only the relevant qualities and relations which are necessary for carrying out the action.
- *Abstraction*. The qualities or relations relevant in the context of the particular action are isolated or stated respectively by an analysis of the action and its elements. This process mostly has the character of a cognitive construction which is directed and regulated by the action. This means that the abstracted qualities and relations are not a characteristic of the objects as such but only apply to them as elements of the action. The theoretical generalizations do not have their root exclusively in the things themselves but in the creating, transforming and operative activity, in the actions of the human being.
- *Objectifying the Symbols*. Thus they lose their original descriptive "character of representatives" and become independent objects as "carriers" of abstracted qualities, relations and elements of the action which on the other hand are denoted and presented by the symbols.
- *Introduction of Variables*. First the concrete symbols have a range of reference which is described by the context of the action, i.e. they are referential variables. The decisive step in the process of generalization is to view these symbols as objectified variables whose characteristic is only given by the abstracted qualities and relations and which stand just for themselves and not for some referents. Thereby the whole system of these qualities and relations becomes a "variable". This variable as a whole has got the character of an object for the subject (this is the abstract-general structure) as well as the substitution property (it can be substituted by different "concrete" structures; it stands for such structures and their common form). In short:

Generalizing means Constructing Variables. In this context a "variable" can be seen as a variable cognitive model (in the cognition of the subject) or also as part of the objective knowledge.

REFERENCES

Bauer, L.: 1984, *Mathematische Fähigkeiten*, Schöningh, Paderborn.

Bender, P. and Schreiber, A.: 1985, *Operative Genese der Geometrie*, Schriftenreihe Didaktik der Mathematik, Bd. 12. Hölder-Pichler-Tempsky, Wien, and Teubner, Stuttgart.

Beth, E.W. and Piaget, J.: 1966, *Mathematical Epistemology and Psychology*, D. Reidel Publ. Co., Dordrecht.

Cassirer, E.: 1980/1910, *Substanzbegriff und Funktionsbegriff*, Wissenschaftliche Buchgesellschaft, Darmstadt.

Damerow, P.: 1982, 'Anmerkungen zum Begriff "Abstrakt" – Philosophiegeschichtliche und mathematikdidaktische Aspekte', in Steiner, H.G. (ed.): *Mathematik, Philosophie, Bildung*. Untersuchungen zum Mathematikunterricht, Bd. 4, Aulis, Köln.

Davis, R.B.: 1984, *Learning Mathematics: The Cognitive Science Approach to Mathematics Education*, Ablex Publishing Corporation, New Jersey.

Dawydow, W.: 1977, *Arten der Verallgemeinerung im Unterricht*, Volk und Wissen, Berlin.

Dörfler, W.: 1977, 'Exemplarische Vermittlung einiger Charakteristika der Mathematik an Hand der Theorie der Matroide', in *Beiträge zum Mathematikunterricht 1977*, Schroedel, Hannover, pp. 63–66.

Dörfler, W.: 1984a, 'Verallgemeinern als zentrale mathematische Fähigkeit, *Journal für Mathematikdidaktik* **5/4**, 239–264.

Dörfler, W.: 1984b, 'Actions as a Means for Acquiring Mathematical Concepts', in *Proc. 8th Internat. Conf. PME*, Sydney, pp. 172–180.

Dörfler, W.: 1986, 'Das Verhältnis mathematischer Operationen und gegenständlicher Handlungen', in *Grundfragen der Entwicklung mathematischer Fähigkeiten*. Untersuchungen zum Mathematikunterricht Bd. 13, Aulis, Köln, pp. 1–14.

Dörfler, W. (ed.): 1988, *Kognitive Aspekte mathematischer Begriffsentwicklung*, Schriftenreihe Didaktik der Mathematik Bd. 16. Hölder-Pichler-Tempsky, Wien, und Teubner, Stuttgart.

Jahnke, H.N. and Otte, M.: 1979, 'Der Zusammenhang von Verallgemeinerung und Gegenstandsbezug beim Beweisen – am Beispiel der Geometrie diskutiert', in Dörfler, W. and Fischer, R. (eds.), *Beweisen im Mathematikunterricht*. Schriftenreihe Didaktik der Mathematik, Bd. 2. Hölder-Pichler-Tempsky, Wien, und Teubner, Stuttgart, pp. 225–242.

Luria, A.R.: 1976, *Cognitive Development – Its Cultural and Social Foundations*, Harvard University Press, Cambridge, Mass.

Morman, Th.: 1981, *Argumentieren, Begründen, Verallgemeinern*, Scriptor, Königstein.

Peschek, W.: 1985, 'Veränderungen kognitiver Strukturen durch Aufbau von Handlungsvorstellungen, in *Empirische Untersuchungen zum Lehren und Lernen von Mathematik*. Schriftenreihe Didaktik der Mathematik, Bd. 10, pp. 215–246. Hölder-Pichler-Tempsky, Wien, und Teubner, Stuttgart.

Peschek, W.: 1986, 'Handeln, Operieren, Denken', *Grundfragen der Entwicklung mathematischer Fähigkeiten*. Untersuchungen zum Mathematikunterricht, Bd. 13, Aulis, Köln, pp. 15–27.

Phillips, D.C.: 1983, 'On Describing a Student's Cognitive Structure', *Educational Psychologist* **18**, No. 59–74.

Pickert, G.: 1984, 'Erzeugung mathematischer Begriffe durch Beweisanalyse, *Journal für Mathematikdidaktik* **5/3**, 167–187.

Rubinstein, S.L.: 1973, *Sein und Bewußtsein*, Akademie Verlag, Berlin.

Welsh, D.J.A.: 1976, *Matroid Theory*, Academic Press, London.

ENTR'ACTE 4

In his introduction to the preceding chapter Dörfler remarks that generalization "can be seen as a psychological process within the cognition of the individual, whose products then are the corresponding cognitive constructs (schemes, frames). But this individual process is always conditioned and mediated socially, as it uses and depends on means attained and prepared by society like language." (p. 63).

This remark is the stepping stone to the next chapter by van Dormolen. When one starts to learn a new field of mathematics one can use the special scientific language that one knows already, but one has also to rely on colloquial language. Take for example the development of the concept of a rectangle as Dörfler describes in Section 3 (p. 75). The student already knows the word 'rectangle' and already has a certain mental scheme for it. Whether or not the individual content of this scheme corresponds with the mathematical content is a matter of speculation. In most cases this would not be true and the teacher and textbook author have to recognise this fact. The 'everyday concept' of a rectangle can, however, be used metaphorically as a means to help students to develop the 'right concept'. Or to say it metaphorically: the student can be helped to draw the everyday concept out of the twilight of her mind.

The next chapter offers a way to deal with this phenomenon of using everyday concepts for the development of mathematical concepts. It brings the reader back to consider the metaphorical nature of learning which was introduced in the first chapter. Van Dormolen uses Searle's view of a metaphor – as a relation between the word, or sentence meaning, and the speaker's meaning. With this model the author analyses difficulties which can occur when metaphors are used in mathematics education, but he also demonstrates that a good understanding of the role of figurative speech in learning concepts helps teachers and textbook writers to use them in their teaching.

JOOP VAN DORMOLEN

METAPHORS MEDIATING THE TEACHING AND UNDERSTANDING OF MATHEMATICS

INTRODUCTION

The subject of this chapter is the use of figurative language in mathematics teaching, in particular the use of metaphors. In this introduction we want to show what place metaphors have in the communication of mathemtatical concepts.

While teaching and learning mathematical concepts teacher and student use professional mathematical speech. This is done in every scientific subject, but, as for example Kemme (1984) showed, professional mathematical speech has special features in comparison to other professional speech.

One of the qualities is the high rate of unambiguousness. The reader or listener (in short the *receiver*) knows exactly what the author or speaker (the *sender*) means. Words and expressions in the professional mathematical speech have, inside the agreed mathematical context, an unambiguous meaning. The sender is not necessarily the teacher, nor the receiver the student. The roles can be reversed. This is meant everywhere in this chapter, unless, of course, the contrary is clear.

Those who are concerned with mathematics teaching know, however, that practice is not so straightforward. In a teaching-learning situation the teacher cannot restrict himself to professional mathematical language.

We do not mean here the situations in which sender and receiver can understand each other very well, when each of them make some effort. In this case the misunderstanding can be caused by thoughtless or even sloppy use of language, or lack of interest and motivation with either the sender or the receiver, and such-like. The more fundamental reason for using other than professional language is because the student does not know the special meaning of all the words and expressions (in short *language signs*), that are under discussion. In these cases it is not possible for many language signs to be transferred to the student by means of a definition.

Take for example words like *associative, axioms, base, even, integral, marginal, odd, paradox, pole, translation, variable.*

Any teacher who wants to teach students a new language sign, can, in certain cases, call upon the knowledge which the student already has. For example, by taking some elements from that knowledge and declaring those as an example of the new concept.

As in: "You can divide 6 by 2 and therefore we call 6 an even number.

Alan J. Bishop et al. (eds.), Mathematical-Knowledge: Its Growth Through Teaching, 89–106.

Also 14 is divisible by 2 and that's why also 14 is called an even number. Any number that is divisible by 2 is called an even number." Or: "These figures are called rectangles".

Often the learning process can be helped by showing non-examples[1] of objects which already exist in the student's knowledge: "You cannot divide 27 by 2 and therefore 27 is not an even number". Or: "These figures are quadrilaterals, but not rectangles". (See, among many others: Ausubel and Robinson, 1986; Van Dormolen, 1974; Van Dormolen, 1981; Freudenthal, 1987; Van Hiele, 1973; Jakobson and Halle, 1956; Pimm, 1987; Skemp, 1979).

As a sender the student will also use, in many cases, language signs which do not belong to professional language, just because he cannot convey his meaning otherwise.

There are many instances in which a sender cannot make himself understandable with signs from professional mathematical language. Then he must call upon non-mathematical language which the receiver understands.

Professional language takes over many signs from non-professional language. These often keep their original meaning, for example in words like *add, figure, horizontal, known, length, less, more, nothing.* There are also signs, and that is the subject of this chapter, which have in the mathematical context of the communication, a meaning which is different from what the receiver knows. Here are some examples of nouns: *angle, degree, diamond, equation, field, function, group, middle, power, root, set, zero*; and of verbs: *divide, draw, express, invert, prove, reflect, rotate, solve, translate.*

In this paper we shall discuss also the figurative use of signs, which will not be part of the professional mathematical speech, but are used to convey the meaning of mathematical ideas to somebody. For example:

- A function is compared to a vending machine.
- Adding and subtracting negative and positive numbers is compared to adding and withdrawing by a witch of cold and hot magical blocks from her magical brew.
- An equation is compared to a pair of scales.

Especially in the cases in which a sign has meaning in the professional language which is different from the meaning, known by the receiver, in the non-professional language, the possibility of misunderstandings between sender and receiver are obvious. Therefore it is worthwile to analyse these situations and to investigate whether this analysis can be applied in practice.

TWO LANGUAGE MEANINGS

A language sign has in the linguistic sense an *internal* (semantic) and an *external* (syntactic) aspect (Keitel *et al.*, 1980).

The external aspect consists of, among other things, the external appearance of the sign. The external appearance, for example, of the equality sign can be described as two equal horizontal lines, one above the other, the length of which is equal to the width of a letter, and placed in the middle of the writing line.

The external appearance of the sign 'apple' is the series of letters a, p, p, l, e in this order. The external aspect contains more, however, than just a certain form. There are also rules, which formally fix the way to use the signs. According to such rules some signs are called 'letter', some combinations of letters are called 'word', some combinations of words are called 'sentence'. There are some rules for the use of letters, words and sentences. Such rules must be known, both to the sender and the receiver, although not always explicitly. The sender and the receiver of this chapter know, that there is no English word with only consonants, and also that to the left and to the right of the equality sign there should be an arithmetical or algebraic expression.

Grammatical rules, or in general, syntax rules, also belong to the external aspect of language signs.

We shall call the meaning, within a certain context, of a language sign, the internal aspect of the sign. The internal aspect of the equality sign consists of something like: "Whatever stands to the right of the sign is as large as whatever stands to the left of it".[2] The internal aspect of the word 'apple' is something like a fruit, which grows on a certain kind of tree, has a certain form, a certain texture, tastes in a certain way, etc. The internal aspect can be very large and can contain much more than can be read from the definition, if a definition exists.

Searle (1979) indicates the internal aspect of language signs by "word or sentence meaning". We want however also to include mathematical signs or expressions, and combinations of words and/or mathematical signs. We shall therefore indicate the internal aspect of a language sign in a more general way by *literal meaning*.

Above we have given some examples in which a receiver learns something new, because he knows the signs transmitted and because he attaches to them the same meaning as the sender intended. Searle (*ibid.*) writes about "speaker's utterance meaning". We do not want, however, to restrict ourselves to spoken language, but also to include written language. Therefore we shall use the expression *intended meaning*.[3]

People use several forms of style in their language to express their meanings. In particular we allude here to the cases in which the literal meaning and the intended meaning do not coincide.

If the sender says: "As the tree, so the fruit", then the two meanings are usually different. The literal meaning of 'fruit' has a biological context, the intended meanings is offspring. The sender expects the receiver to understand the intended meaning.

We shall depict the relations between intended meaning and literal meaning by pictures. We shall do it in the way Searle (*ibid.*) did it. If intended meaning and literal meaning do not coincide we depict this by

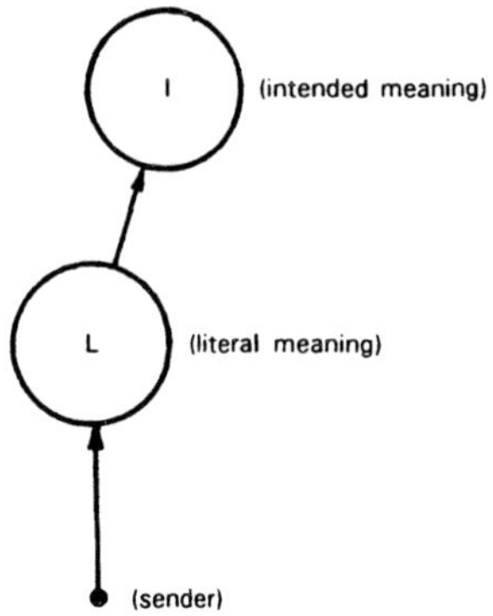

Both arrows symbolize an activity of the receiver. The lower arrow points to the registration of the signal by the receiver. The upper one to the search of possible intended meaning.

For example:

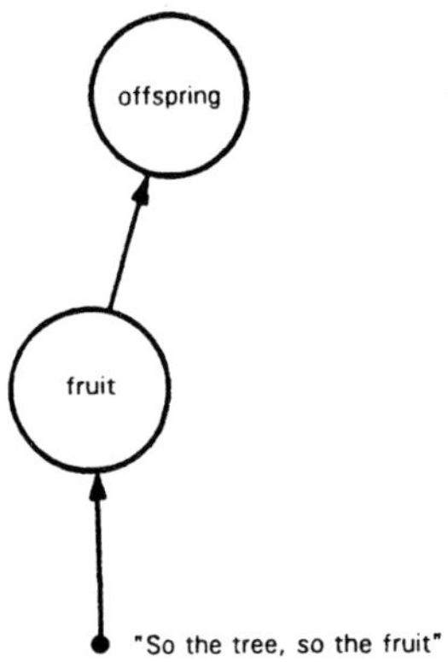

If the intended meaning and the literal meaning coincide, we shall depict this by

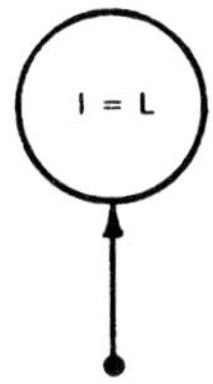

For example

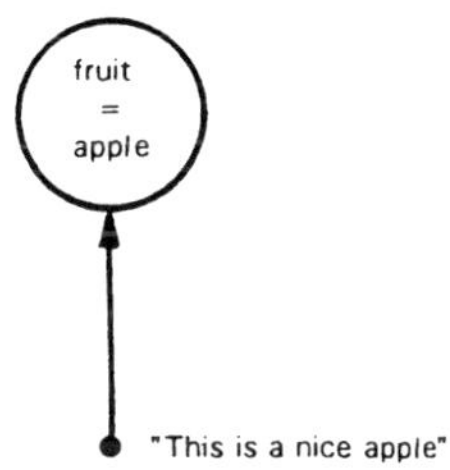

In the first case ($I \neq L$) the receiver understands the figuratively meant text, in the second case ($I = L$) the literally meant text is understood as such.

A PROBLEM OF COMMUNICATION

With help of the diagrams we can illustrate a problem in communication. There are several possibilities in the sending and the receiving of language signs.

a. The receiver interprets the transmitted signals differently from the intended meaning.

a1. One possibility is when the receiver interprets the signals literally, while this is not meant to be the case ($I \neq L$).

Example: Once a student told her teacher, that she did not agree that one can draw only one line through two points. To explain this she drew two points on a piece of paper, put a ruler along them and drew her pencil several times along the ruler. The teacher's intention with "one can draw only one line" was meant figuratively, while the intended meaning was "there exists only one line".

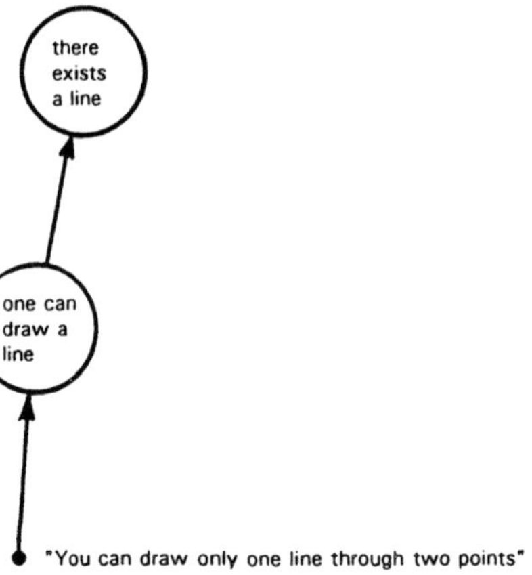

The student used the literal meaning.

a2. Another possibility is that the sender meant the signal to be interpreted literally ($I = L$), while the receiver takes it figuratively.

Example (illustration on next page): In Dutch the word 'assign to' ('toevoegen') can mean to mark something or someone as belonging to something or someone else (like assigning an officer as an aide to a general); but it is also often used as adding some material to something else (like adding sugar to the coffee). The last meaning is more commonly used. As a result students often misinterpret the definition of a function, when in that definition the word 'assign' is used. For example: The teacher gives the function which assigns to a certain number its square ($x \rightarrow x^2$), then some students assign 2 to 1, 6 to 2, 12 to 3, etc.: They calculate the square and add the result to the original number.

a3. A third possibility is that the receiver understands that the literal meaning and the intended meaning do not coincide ($I \neq L$), but assigns another meaning to the signals from that which was intended by the sender. Example: In a text book is written:

"(a) Patrick has a score of 3 points on Wednesday night and a score of 2 points on Thursday night. What did he score on Thursday?

(b) The situation in (a) can be written as an equation:

$$3 + ? = 2.$$

What is the meaning of ? in this story?" (Van Streun, 1986).

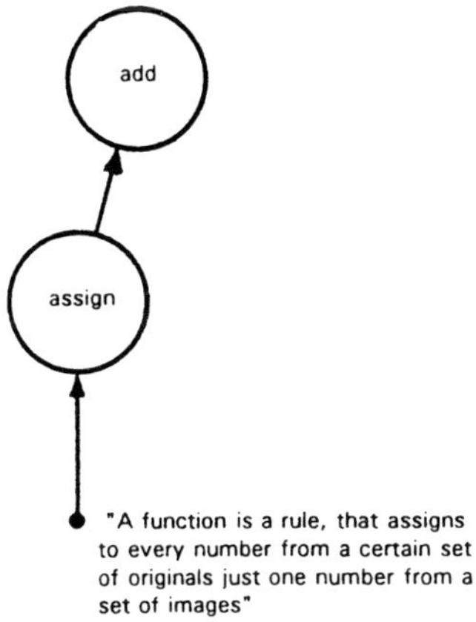

The literal meaning of a question mark is something like: The sentence, written before this, is meant as a question. In the equation in the text there is no sentence before the question mark. This gives the reader a signal that the question mark must be meant figuratively. The intended meaning is: What number can you write here, so that you get a true assertion?

A student wrote in his copy book: "? = he did nothing special or they do not know what he did".

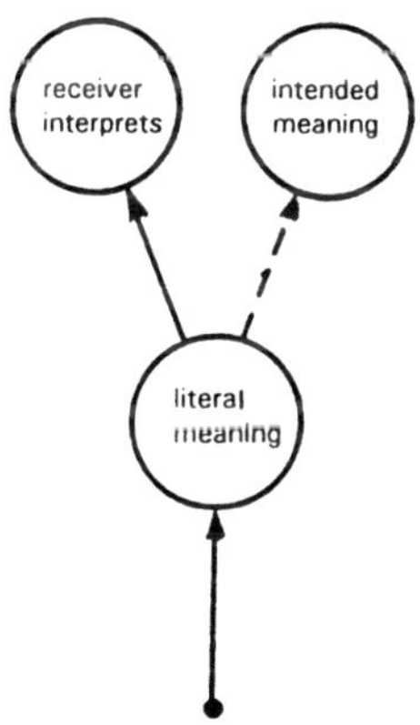

b. The receiver does not understand how the transmitted signals relate to the context. In the intended context they have no meaning for the receiver.

It is remarkable that students can often give the right answers to exercises in situations like this. They have learned to imitate examples of such exercises. When one asks them how they got to the answer, they reply something like: "It has to be done this way".

c. The receiver understands the signals just as they are intended, either literally or figuratively.

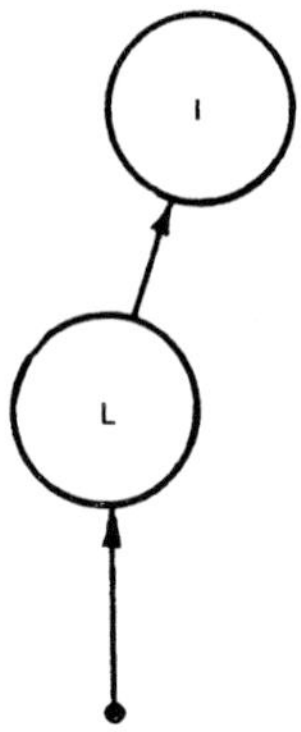

DIFFERENT KINDS OF FIGURATIVE SPEECH

We shall not enumerate all the possibilities of the figurative use of language. In this chapter we are especially interested in the two figures of speech called *metaphor* and *metonymy*.

As a sender we can often warn the receiver in advance that language will be used figuratively. In the choice of our words we indicate that the receiver must look for a meaning which does not coincide with the literal meaning.

Examples:
"*As the proverb says*: As the tree, so the fruit",
"*You should imagine* numbers as points on a line",
"A function *is something like* a vending machine",
"*Think* of a vector *as* an arrow",
"*It looks like* he is sitting on his money, because he is so thrifty".

In such cases there is a question of an analogy. One could understand an analogy as the comparison of two relationships:
$A : B \quad C : D$ (Miller, 1979; Pimm, 1981). The sender does not make explicit which relations are meant and in which way the comparison is done.
"A tree is to its fruit, as the parent to his offspring",
"Just as the points are next to each other on a line, so is the succession of numbers",
"Just like you can get something from a vending machine by putting money

in the slot, so you get a number from a function whenever you put a number in".[4]

We leave the translation of the other examples to the reader.

Bauersfeld and Zawadowski (1981) take a wider stand. With an analogy, two situations are compared, and with it is suggested that both have a similar structure. The sender keeps implicit how the structures are. This has to be so, because if the structure could be described explicitly, the figure of speech would not have been necessary, unless for esthetic reasons. The power of undefined analogies lies in the fact that people are able to learn through such analogies (Hofstadter, 1986).

When one does not say what structure one means exactly, one does not need to convey the message necessarily in the form of an analogy. One can leave out parts of the analogy. Then one gets the metaphoric figure of speech.

Examples:
"Numbers are points on a line",
"A function is a vending machine",
"A vector is an arrow",
"He is sitting on his money",
"-1 is an imaginary number",
"Night is falling",
"John is a pig",
"You cannot draw more than one line through two points",
"3 + ? = 2".

In all of these cases the metaphors can be restored to an analogy. It should be clear that the sender strongly calls on the receiver. The latter receives the message and registers that the literal meaning L cannot be meant here. This would conflict with the context. From this anomaly the receiver concludes that it is necessary to look for a special meaning I. By intentionally including an anomaly the sender warns, but this time implicitly, that an intended meaning I ($\neq L$) should be looked for.

Black (1972) has shown that the idea that a metaphor can be restored to analogy ("comparison view") is insufficient to study in depth the phenomenon of metaphor. He pleads for a view in which the literal meaning and the intended meaning influence each other ("interaction view"). In "John is a pig", for example, 'John' does not convey the same meaning after transmission and receiving of the sentence as before. But also 'pig' has got another interpretation. The word no longer stands for a certain animal, with certain biological qualities, but now has certain stereotyped meanings. It depends on what the sender and the receiver know about John, and about each other, whether the pig is now a very muddy animal, an animal that eats noisily and messy, or is a beast with sexual aberrations.

In particular for situations which cannot be described literally, the comparison view plays an important role. This is also the case in which someone has to convey to others a mathematical concept, which the others do not know yet. In particular this is the case when the concepts under discussion are of an elementary nature, which can only be described by axioms or by very abstract definitions, such as point, number, variable, set, quadrilateral, angle, function, etc.

Another figure of speech is *metonymy*. This is a form in which the intended meaning cannot be deduced from the literal meaning by a comparison of structures, but by looking for some cohesion between the literal meaning and the intended meaning.

Examples:
"I had a telephone call from *New York*",
"There were 40 *heads* on the ship",
"Did you take my *Shakespeare*?",
"*Area* equals *length* times *width*",
"Euler's theorem for polyhedrons is $V - E + S = 2$".

Like metaphors metonymies are also important means to convey concepts through language. In a metaphor I and L have more or less the same (suggested by the sender) structure, but they differ in content. The figurative meaning is founded on the comparison of the structure of I and L.

In a metonymy I and L have the same content, but they differ in structure. The figurative meaning is founded on the cohesion between I and L.

Metaphors and metonymies are complementary to each other, as Jakobson and Halle (1956) have shown with people who have the speech disturbance aphasia.

Other forms of figurative speech, such as indirect speech, or ironic speech (Searle, *ibid.*) will not be analyzed here.

DEAD METAPHOR

Sometimes figurative speech has become a set expression. It is no longer a metaphor. One could speak of a dead metaphor then.

An expression like "Twilight falls" is originally a metaphor. I only realized this since I am in close contact with persons who have a mother tongue which is different from mine. One of them reacted surprisingly when he heard the expression: "How can twilight fall?" The sender did not realize that he did use "to fall" in the literal meaning.

Such situations happen repeatedly in education. Mathematics teachers in the Netherlands know the surprised looks in the students' eyes when they hear for the first time about "*cutting* lines", which is in the Dutch language the standard expression for intersecting lines. One can imagine reactions like these from English speaking children, when they hear about lines 'that *meet*'.

In Dutch the word for rhombus is 'ruit', which is the same word as for a glass pane. American children might not be surprised when they hear the word diamond for rhombus, because this is the word they know for the figure in the baseball field.

Other examples, in which misunderstanding is possible, are:

- Number *line.*
- The *square root* of...
- The *limit* of...
- *Real, imaginary* and *complex* numbers.

We expect some problems in English with the 'square of a number'. In Dutch this is not so, because in Dutch the words for the geometric object and the arithmetical object are different.

Confusion is possible when the teacher or the author of the student text uses words which have a meaning (intended meaning) in the mathematical context which is different from the meaning (literal meaning) in the general context, and does not realize that the children do not know the mathematical meaning.

When the sender realizes this, he can warn his receivers by using expressions like: "This we call...", or "In mathematics, this is called ...". This does not solve the problem however, that the intended meaning is completely new for the receiver and is, consequently, not absolutely integrated into his own knowledge. In some cases the receiver realizes the anomaly ("What have numbers to do with lines?", "How can a line cut, it's not a knife, is it?") and he can search in his mind for the intended meaning. In most cases the receiver does not realize the anomaly and the search takes place unconsciously. In neither case however can the receiver find the intended meaning, because this does not belong to his knowledge yet. We can depict this situation by:

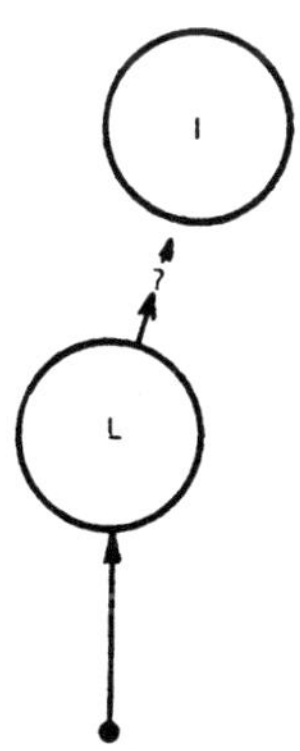

The receiver cannot find *I*, because he cannot find, in the current context, a structure which is comparable with the structure of *L*.

For example, to understand the analogy with the literal meaning of 'root' of a number, one should know how this word was introduced in professional mathematical language (e.g. Van de Waerden, 1985, 1961). Those who cannot get the feeling of problems like these, should try to remember their own learning problems when they heard or read for the first time words like 'field', 'ring', 'filter', 'ideal', 'imaginary', 'determinant', 'fraction', 'discriminant', 'adding of vectors', etc.

The sender must realize that the use of dead metaphors can make problems for the receiver, because the latter lets the literal meaning coincide with the intended meaning, or searches in vain for the intended meaning which is appropriate in the given situation.

METAPHORS ARE INDISPENSABLE FOR LEARNING.

In the following we shall investigate some examples from mathematics education in which metaphors are important for the learning of mathematical concepts in particular.

a. Metaphors for the Equality Sign

Formally the equality sign is the symbol for the equality of two expressions.[5] In practice there are more meanings:

In "$(a\text{-}3)(a+5) = \ldots$" it has a signal meaning: the student is urged to undertake an activity,[6] in this case to change the expression to an expression without brackets.

In "$(a\text{-}3)(a+5) = 0$" it has also a signal meaning, but a different one from the former. The student is urged to solve the equation. It also, hopefully, has a symbol meaning: Here are some numbers which make the sentence true when they are substituted for the variable a.

In "$y = 3x - 5$" and in "$y + 5 = 3x$" the equality signs can have different symbol meanings. In the first for many students the meaning is something like: "Whenever you know a number for x, you can calculate the corresponding number for y". In the second equation x and y have a much more symmetrical role.

These examples make it clear again that the meaning of a symbol is subjective, and depends upon the context.

b. Metaphors for the Concept of Limit

The concept of limit arose from the concept of motion, which is time-dependent. Kaput (1979) writes: Perhaps the most obvious is the motion metaphor that gives

$$\lim_{x \to a} f(x) = L$$

it primary meaning. "As x moves towards, $f(x)$ moves towards L". This is frequently written:

$$\text{as } x \to a, \quad \text{then } f(x) \to L.$$

Nowhere is the schizoid relationship of mathematics more blatant than here.

The formal mathematical meaning however, laid down in the epsilon-delta definition is timeless. Students must struggle to ban the time element from the meaning.

c. Metaphors for the Minus Sign

The minus sign occurs in three different meanings: a sign for position, as in -3; a sign for an operation as in 5-3; and a sign for the opposite, as in $-a$ (e.g. Wansink, 1976). Metaphors are indispensable for teaching students who learned only the second meaning in primary school, the two other meanings. Several meanings are in use. It is worthwhile investigating in what way their use can block future developments, such as the metaphor of the witch with cold magical blocks (negative numbers) and hot ones (positive numbers) (De Vos, 1986).

d. Metaphors for the Concept of Variable

The use of a letter is originally a metaphor. Teachers know the problems which arise in the learning of the concept (Van Dormolen, 1984). Here also several methods are used to introduce the concept through metaphors, such as open place, open squares, question marks.

Metaphors are important as didactical means for the teaching of certain concepts. One can even say, that the use of metaphors is indispensable (Hofstadter, 1986; Pimm, 1981; Kaput, 1979). The reverse side of the medal is the danger that students are bound to them in such a way that they block further learning. In the following we will show, that these circumstances can be analyzed with Van Hiele's theory of levels of understanding (Van Hiele, 1973, 1986). We are convinced that the use of metaphors forms a transition phase between the ground level, in which the students let the literal and the intended meaning coincide, and the first level, in which the literal meaning acts as a model. In the latter case one can speak of a dead metaphor.

METAPHORS IN THE TWILIGHT ZONE

Teachers and text book authors know how difficult it is, if not impossible in most cases, to get students to understand a new concept by way of a formal description.[7] In these cases the use of examples and/or figurative speech is indispensable.

Here lies the seed of a problem of misunderstanding. It is not always possible to use examples from a mathematical context. There are several situations in which this would be the case. One of them is the situation in which the new concept is of a higher order than the concepts which the students already know (Ausubel, 1969; Skemp, 1971).

Another situation arises when the teacher (or author) wants to convey a concept of a very elementary nature, like 'point', 'line', 'number'. In this case there is no possibility to refer to mathematical background knowledge, which would be about axioms and logical rules, and therefore of a very high level nature.

In those cases the examples must come from non-mathematical contexts, and because of that these examples play a metaphorical role.

Here are some such situations.

- Children learn the number concept from counting. There is a strong connection between number and amount. But we cannot show examples of numbers, or for that matter of amount. We have to show a certain amount of marbles, of flowers, of people, of motorcars, etc. Then we must hope the child learns to abstract (or generalize) the amount from the sets it is counting. At a certain moment the marbles, flowers, people, motorcars are not important any more. They even have to be discarded as totally irrelevant.
- The subtraction of negative numbers only has examples which are very difficult to understand within the mathematical context. It can be made plausible by series like

$$6 - 4 = 2$$
$$6 - 3 = 3$$
$$6 - 2 = 4$$
$$6 - 1 = 5$$
$$6 - 0 = 6$$
$$6 - -1 = \ldots ? \ldots$$

It looks extremely plausible to say that $6 - -1$ should be 7. But this is not an explanation. Children understand the pattern and therefore accept as an algorithm that 'subtraction with a negative number is the same as adding with its opposite'. They even say that they understand, but they understand how it is to be done, not why it is to be done in this way.

We have seen another attempt with examples from within the mathematical context. A teacher explained, using the number line:

'6 - 4 brings you 4 to the left of 6. Now -4 is the opposite of 4, so it is reasonable to say that 6 - 4 brings you 4 to the opposite side, that is the right of 6'.

In many cases we hear (or read in texts books) attempts to use non-mathematical situations. A famous one (or should we say notorious?) is the taking away of debts. Children seem to understand very well that taking away a debt leaves one without debt. It is hard to understand however, how one can take away a debt of 10, when one has a debt of only 4.

Van Streun *et al.* (1986) use a description of a boarding school in which students can get honorary points for good behaviour and penalty points for bad behaviour. Teachers can give you a certain number of honorary or penalty points. They can also take them away.

A teacher used this to explain adding and subtracting positive and negative numbers. The adding gave no difficulties: Apparently it was clear to the students, that if one has 6 penalty points and one gets 8 honorary points, the result would be 2 honorary points: (-6) + (+8) + (+2).

For some of the students however, 'taking away' was only a metaphor for subtracting if the amount that was taken away did not exceed the original amount. So, the example (-6) - (-4) = (-2) was well understood. In the other case, such as (-6) - (-8) some students had difficulties. As one student said: 'How can you take away eight penalty points, if you only have four of them'? A friend said: 'Of course you can, this is only doing as if'.

The first student took the figurative speech literally. Her friend appeared to understand the intended meaning. She could not, however, tell immediately the result of (-6) - (-8). She needed to translate it first to penalty points and taking away. But then she had no trouble with any number. Other students did not need any translation.[8] They could add and subtract any positive and negative numbers formally, without the use of any non-mathematical aids.

Many authors have described the different levels of use of language (e.g. Ausubel c.s. 1969; Freudenthal 1978; Van Hiele, 1973). We want to concentrate here on what Van Hiele calls the ground level or zero level and the first level. It describes the difference between the situation in which the learner is still bound by the examples as concrete objects and the situations in which it is clear for the learner that all examples are in a certain sense the same.

We contend that there is *a stage between these two levels*, in which the learner already knows that the examples as such are irrelevant, but is not able yet to discard them. The examples are still necessary for the thinking process. This is the stage in which the learner sees the examples as metaphors.

At the ground level the examples are not examples at all. The teacher calls them examples, and by this the students understand that there will come a stage that they will know more about it. At the moment, however, they are independent situations. Each new instance is a new situation. For them the only meaning of the language is the literal meaning, like the girl who could not 'take away' minus eight from minus six.

In the twilight zone between the ground level and the first level, the students understand that they are working with examples of some concept. They can operate with these examples metaphorically. That means, they can separate the intended meaning *I* from the literal meaning *L*. This was apparently the case with the girl who translated subtracting to 'taking away', but then could take away anything, even if in ordinary life such a thing would be preposterous.

At the first level of abstraction, the students see that all the situations are instances of the same situation. They do not need the examples any more. They can operate in a formal way. If ever there was a metaphor, then it is now a dead one.

Also the case of understanding 6 - -4 as 'going to the opposite side of 6' is a good example of this metaphoric twilight zone. It is the situation in which the teacher expected (although she was not aware of this) that her pupils would not take 'going' literally in the physical sense. She used the word however because she expected that she would lose them if she used formalized language.

NOTES

[1] Non-examples are not the same as counter examples. Counter examples are special cases that do not fit a certain assertion. Non-examples are special cases, that are almost, but not really, examples of a certain concept.

[2] We shall not try to define the internal aspect of the equality sign unambiguously. The case is even more complex. For some people the internal aspect of the sign is different. It looks more like an order: "Compute what stands to the left of the sign and write the result to the left", or: "When you compute what stands to the left of the sign, you should get what stands to the right of it". For such persons

$$(a+b)(a-b) = a^2 - b^2$$

is something different from

$$a^2 - b^2 = (a+b)(a-b).$$

[3] Hofstadter uses a nice alliteration for both kinds of meanings: "literary and literal meaning". It was tempting to adopt this, but we resisted the temptation, because we wanted to stress the role of the sender and the receiver.

[4] This is a much too narrow interpretation of function, of course. We mention it here only to illuminate our meaning.

[5] The inventor of the equality sign, Robert Recorde (Whetstone of Witte, 1557), also used a metaphor when he introduced the sign as two lines of about 2 cm long, just under each other: ", bicause noe .2. thynges, can be moare equalle." (e.g. Cajori, 1952).

[6] A certain sign can have a symbol meaning and a signal meaning. We differentiate these two because in the first case the meaning of the sign refers to a certain concept in the person's mind. In the second case the symbol urges the reader or listener to undertake

some activity (Van Dormolen, 1982). It is a well known fact that many students are confused when they see something like 5 + 3 = 2 + 6. They do not accept what is written to the right of the sign. There should be an 8. Those students do not assign a symbol meaning to the equality sign, while the signal meaning is something like: "calculate what stands to the left and write the result to the right". For such students it is completely acceptable to write the solution to "Calculate 2 + 5 + 6 - 4" as follows: "2 + 5 = 7 + 6 = 13 - 4 = 9".

7 The word 'concept' is meant here in a very broad sense. It refers also, for example, to a new way of working, to a new algorithm, to a new theorem, to a new convention.

8 We did not observe the students from the beginning. The observation that is reported here, was done about half a year after the learning of subtracting negative numbers. The teacher wanted to freshen up her students' memory.

REFERENCES

Ausubel, D.P. and Robinson, F.G.: 1969, *School Learning, an Introduction to Educational Psychology*, New York.

Bauersfeld, H. and Zawadowski, W.: 1981, *Metaphors and Metonymies in the Teaching of Mathematics*, Occasional paper No. 11, IDM, Universität Bielefeld.

Black, M.: 1972, *Models and Metaphors*, London.

Cajori, F.: 1951–1952, *History of Mathematical Notation*, 2 vols., Berlin.

Davis, R.B.: 1975, 'Cognitive Processes Involved in Solving Simple Algebraic Equations', in *The Journal of Children's Mathematical Behavior* **1**, vol. 3, 7–35.

Dormolen, J. van: 1977, 'Learning to Understand what Giving a Proof Really Means', *Educational Studies in Mathematics* **8**.

Dormolen, J. van: 1981, *Didactiek van de wiskunde*, Utrecht.

Dormolen, J. van: 1982, *Aandachtspunten, de a priori analyse van leerteksten voor wiskunde bij het voortgezet onderwijs*, Utrecht.

Dormolen, J. van: 1974, 'Variabelen', in: *Euclides* **60**, No. 3, November 1984. Freudenthal, H., *Mathematics as an Educational Task*, D. Reidel Publ. Co., Dordrecht.

Freudenthal, H.: 1978, *Weeding and Sowing*, Dordrecht.

Hiele, P.M. van: 1973, *Begrip en inzicht*, Purmerend.

Hiele, P.M. van: 1987, *Structure and insight*, New York.

Hofstadter, D.R.: 1986, *Metamagical Themas: Questing for the Essence of Mind and Pattern*, New York.

Jakobson, R. and Halle, M.: 1956, *Fundamentals of Language*, Den Haag.

Kaput:, 1979, 'Mathematics and Learning: Roots of Epistemological Status', in Lockhead, J. and Clement, J. (eds.), *Cognitive Process Instruction*, Philadelphia.

Keitel, C., Otte, M. and Seeger,F.: 1980, *Text, Wissen, Tätigkeit*, Königstein-Ts.

Kemme, S.: 1984, *Taalaspekten van het wiskundeonderwijs*, Vakgroep OW&OC, Utrecht.

Miller, G.A.: 1976, 'Images and Models, Similes and Metaphors', in Mooij, J.J.A. *A Study of Metaphor*, Amsterdam.

Ortony, A.: 1979, *Metaphor and Thought*, Cambridge.

Pimm, D.: 1981, 'Metaphor and Analogy in Mathematics', in *For the Learning of Mathematics* **1** No. 3, 47–50.

Pimm, D.: 1987, *Speaking Mathematically, Communcation in Mathematics Classrooms*, London & New York.

Searle, J.R.: 1979, 'Metaphor', in Ortony, A., *Metaphor and Thought*, Cambridge.

Skemp, R.R.: 1971, *The Psychology of Learning Mathematics*, Worthington.

Steiner, H.-G: 1988 Über Metaphor, Modelle und Mathematik', in Bender, P. (Hrsg.),

Mathematikdidaktik, Theorie und Praxis, Berlin.
Streun, A. van *et al.*: 1986, *Wiskunde lijn*, Vol. 1a, Groningen.
Vollrath, H.-J.: 1978, *Lernschwierigkeiten, die sich aus dem umgangssprächlichen Verständnis geometrischer Begriffe ergeben*, Schriftenreihe des IDM 18, Bielefeld.
Vos, W. de: 1986, 'Het belang van vakdidactisch onderzoek voor een samenhangend onderwijsbeleid', in *Tijdschrift voor de didactiek der beta-wetenschappen*, Vol. 3, No. 4.
Waerden, B.L. van der: 1961, *Science Awakening*, Groningen.
Waerden, B.L. van der: 1985, *History of Algebra: from al-Khwarizmi to Emmy Noether*, Berlin.
Wansink, Joh.H.: 1967, *Didactische oriëntatie voor wiskundeleraren*, Groningen.

ENTR'ACTE 5

The chapter by van Dormolen demonstrates to the reader the range and variety of metaphorical representatives of mathematical knowledge which are available to the teacher, and ultimately to the learner as well. One task for the teacher is to learn how to make decisions about the use of these metaphors in helping the learners gain access to mathematical knowledge. In the next chapter Douady takes the reader one level further by considering a particularly powerful set of metaphors and their implications for teachers' decisions. These metaphors are elements and ideas which can be used to interpret the didactical situation. One complementary pair of metaphors is 'mathematics as a tool' and 'mathematics as an object' and Douady explores the relationship between these two metaphors and also the didactical tension provoked by their dialectic. The other elements in her analysis are 'setting' and 'window'. Central to the discussion are the questions: what does it mean to 'know' mathematics, and how can the teacher conceptualise the relationship between the teacher's mathematical knowledge and that of her pupils?

RÉGINE DOUADY

TOOL, OBJECT, SETTING, WINDOW: ELEMENTS FOR ANALYSING AND CONSTRUCTING DIDACTICAL SITUATIONS IN MATHEMATICS

INTRODUCTION

The classroom is a living place where complicated interactions take place between the teacher and the pupils. What is at stake is a certain mathematical knowledge. The pupils arrive in class in a certain state of knowledge and must reach another expected state of knowledge, under the teacher's direction. Various factors act upon these relations – of a scientific, social, cultural, hierarchical, or personal order – often in contradictory ways. One of the functions of teacher training is to provide teachers with the means of recognizing those factors which they can influence, considering the constraints to which they are subject. Another point is to determine how they can manage these elements in order to obtain a desired result in the pupils' learning.

In our view, however, the analysis of the constraints and of the margin of manoeuvre, and therefore the possibilities of action, the decisions and the possible realization in class depend significantly upon the teacher's *representations* concerning mathematics and the mathematical activities (not just scientifically but also from a cultural or social viewpoint).

We present in Section 1 a set of questions concerning the relationship between mathematics teaching and mathematics learning in the school (or university) situation and concerning the metacognitive representations of the teachers. In Section 2 we state some epistemological elements to analyze the mathematical knowledge and to determine our choices, and in Section 3 their didactical consequences in terms of a model for the global organization of the teaching-learning process.

1. SOME QUESTIONS AND REFLECTIONS

What does 'to know mathematics' mean? What are the constraints imposed on the teachers and what is the area of freedom they can use in order to fulfill correctly their office? How should they manage such areas of freedom?

For a certain epistemological position and a given formal knowledge to be taught, how can it be subdivided and put "on stage" in order to be treated

Alan J. Bishop et al. (eds.), Mathematical Knowledge: Its Growth Through Teaching, 109–130.

meaningfully by pupils? Are the means of access to this knowledge the same for all pupils? for all pieces of knowledge?

How would the teachers organize and conduct the teaching so that the observed effects are as close as possible to the expected effects? What are the means of regulation which the teachers use in the negotiation with the pupils? What kind of knowledge is at stake in this negotiation? How would they analyze and interpret the possible discrepancies between what is found and what is expected?

Under which conditions is it possible to integrate the cognitive, social and cultural diversities of the pupils in a collective process?

Actually, the questions above involve the representations of the teacher concerning what mathematics is, what mathematical activity is, what its role in the school curriculum is, what the place of the teacher is in the learning process. These *metacognitive* representations are influenced by the knowledge they may have of mathematical life, possibly through history, research, reading texts or simply their experience as a pupil.

From their professional experience, teachers are led to distinguish external data and explicit constraints (knowledge acquired by pupils out of school, curriculum, timetable, number of pupils...), pressures (expectations of parents, of school inspectors, of the pupils themselves) and the prevailing educational customs. These pressures are not only of a scientific order, but also of a social or an ideological order.

Internal pressures (the representations) combined with external pressures condition choices and decisions concerning access to knowledge, concerning expectations and requirements for the pupils and the ways of taking into account their personality and their diversity in a class. These pressures filter the instructions and comments coming from the institution or the schoolbooks concerning the content of the teaching, its arrangement and its precise presentation to pupils, concerning time and space management.

With respect only to the scientific dimension, a teacher who believes that it is necessary for the pupils to know the definitions, methods and theorems needed for a certain problem before they tackle it, will hesitate before proposing to his pupils a problem when they don't have the operative means for its solution. If he thinks that learning is above all a personal business, he will be reluctant to have his pupils work in groups. If he thinks that what is important is a certain familiarity with properly labeled skills, or if the conceptualization work which justifies them seems to be too costly, he will be tempted to skip the work required for the pupils' conceptualization. He may do so even if he is convinced that such work would increase the pupils' ability to adapt their knowledge in unusual situations.

On the other hand, the representations of the pupils themselves concerning mathematics, mathematical activity, its usefulness, the way they can learn mathematics, and so on, interfere with those of the teacher and influence their own way of responding to the teacher's expectations.

Concerning the learning, we make the hypothesis that beyond the diversity of pupils and situations, there are regularities in the learning process in school, that it is possible to build a teaching strategy by taking them into account, and that such teaching improves the efficiency of acquisition for a number of pupils.

Didactical research on these topics is of interest to all teachers. It may help them to detect and to understand class phenomena when they occur. It may help them to organize their teaching and predict certain results, to confront the reality with expectation. It may help to facilitate their rational decision-making.

2. SOME EPISTEMOLOGICAL ELEMENTS

In order to define our position, we have taken into account phenomena during which mathematical knowledge is born and developed, as well as explanatory hypotheses and conditions which may induce them.

To do that, we introduce the following notions: *tool, object, setting, conceptual window* and the following relations: *tool-object dialectic, changing of settings, interplays between different settings, subdivision of the mathematical field into conceptual windows.*

We shall describe these notions and relations. We describe just below what we mean by mathematical concept.

2.1. Beings and Concepts in Mathematics

The elaboration of concepts is a process by which the human being becomes aware of reality, a process by which he gives preference, in those concrete situations which he wants to analyze, to some aspect or characteristic. Then, he can gather according to that characteristic various exemplars and make new categories, oppose some characteristics to others, and establish relations between them. While rooted in an observable context which provides meaning to them, these characteristics free themselves of it and acquire autonomy, a life of their own. They are concepts.

All phenomena dealt with by Mathematics have, like concepts, an ideal nature. Some are brought forth by the modelling contexts outside mathematics, such as real numbers, for instance to measure quantities; others such as complex numbers are produced from purely theoretical considerations to solve problems inside mathematics, some to fit the needs of generalisation or unification, such as Linear Algebra. In the mathematical life, some well-established phenomena such as numbers or geometrical objects play in turn the role of concrete individuals about which new problems are being set, including modelling, that release new invariants (for instance preservation of length, area, volume by translation or rotation) between which relations will be established, inducing the birth of new mathematical phenomena (such as isometries), new relations which have different properties. Organizing properties or structures are the objects of study for the mathematician. By this means is established in the mathematical world a dynamic of fundamental questions and problems creating objects and relations. We refer to them as *mathematical concepts*.

2.2. Tool and Object, Changing of Settings Through Examples

During their evolution and adult life the mathematical phenomena receive various statuses. Two of these statuses, denoted 'tool' and 'object', are involved in a strong and differentiated manner in the epistemological analysis of a mathematical notion, and also in the study of the relation between teaching and learning. Taking them into account may become a decision element for the teacher.

In order to explain what we mean by tool and object, we shall describe an example arising from the history of mathematics and another one from the teaching of mathematics at school. Then we shall give a general formulation.

Example 1: The resolution by radicals of third degree equations

In the 16th century, Tartaglia mentions to Cardan the resolution technique of S. del Ferro. It relies on the identity $(x+y)^3 = x^3 + y^3 + 3xy(x+y)$. This identity is an *explicit tool used by somebody in a definite context.*

Let us consider an equation of the form $x^3 + px + q = 0$.

S. del Ferro's technique provides for the solution an expression which involves $4p^3 + 27q^2$. Cardan notices (Ars Magna, 1545) that this technique cannot be used – staying with real numbers – in the case where $4p^3 + 27q^2 < 0$ since either a positive or negative real number has a positive square. In other words, the technique fails where there is more than one solution. On the other hand he can exhibit equations with $4p^3 + 27q^2 < 0$, having 3 real roots. Paradox! There is an imbalance between what he expects in the matter of equations and what he can do with real numbers.

In order still to be able to apply the technique, the idea comes up that one has to leave the real numbers. The necessity of building a new setting grows.

A few years later, Bombelli solves the problem by *introducing two new signs* "piu di meno" and "meno di meno" and proposing *a sign rule* for the multiplication which provides the set {+,-,+ di-, -di-} (implicitly at that time) with what we would now call a group structure. The rules are validated by the good results they produce in terms of real solutions. Computation involving square roots of negative numbers (called *imaginary numbers*) are developed.

The concept of complex number is being slowly matured: implicit tool, support of explicit techniques and practices with symbols in the 16th century, the status of imaginary numbers will evolve in the 17th and 18th centuries.

A. Girard, in "l'invention nouvelle en Algèbre (1629) states the *permanence principle* which asserts that any identity obtained with real numbers can be applied with imaginary quantities.

Descartes asserts a *principle* "every equation of degree n admits exactly n solutions, some of which are possibly imaginary". This statement gives weight to the idea that imaginary quantities may have the status of number.

Those imaginary numbers began to enter into the reasoning. The extension of functions to imaginary values is considered for polynomials, for series (without worrying about convergence). In order to integrate rational fractions, a method appears: decompose them into simple elements. The case where such a decomposition involves imaginary poles induces the necessity of giving a meaning to the integral of those expressions, and thus to extend to imaginary quantities the integration of real functions.

The contexts and the settings involving imaginary quantities are enlarged.

However the inducement to give them the status of number encourages the process of dealing with them like real numbers. The extension to imaginary quantities of what is usually done with real numbers provokes, at the beginning of the 18th century and for several years, a *controversy* between J. Bernoulli and Leibniz *concerning the logarithms of negative or imaginary numbers*. The implicit hypothesis of the two protagonists is that the logarithmic function extended to imaginary quantities exists per se. Euler makes it explicit and denounces it. On this occasion, the knowings about imaginary quantities improve. But they are still not constructed. Practices are developed concerning them.

Concerning Descartes principle, it is disputed by N. Bernoulli, supported by Euler, *d'Alembert* "proves" it *in 1746* and even "proves" that *all imaginary numbers – by this, he means all roots of polynomial equations – can be expressed in the form* $a+b\sqrt{-1}$, where a and b are real numbers. D'Alembert considers them for themselves, independent of the context of use. *Imaginary numbers begin to acquire the status of object.* But there is still a long way to go.

At the dawn of the 19th century, the imaginary numbers still do not have the status of a mathematical and cultural object, independent of the context in which they are used and of the convictions of the mathematicians who handle them. The works of mathematicians in the 19th century, mainly Gauss and Cauchy, lead to awarding complex numbers the status of *object*.

The complex numbers are defined at least in three different settings. In the geometrical setting, we mention the contributions of Wessel (1798) and Argand (1806), but it is Gauss who first provides a complete exposition of this question, and Cauchy who popularized it. In the arithmetical setting we have to consider the contribution of Hamilton. In the algebraic setting, the important contribution is that of Cauchy.

In problems, these three approaches often interact.

Example 2 taken from the class
(French level 3ème – 2ème, age 15–16).

The pupils concerned are familiar with some algebraic rules, with developed expressions. They know how to factorize in easy cases at the request of the teacher.

Problème: Solve the following equation $x^2 - 4 + (x + 2)(2x - 5) = 0$.

The term "equation" is an object of the algebraic setting. The expression $x^2 - 4 + (x + 2)(2x - 5) = 0$, a degree 2 polynomial, is another object of the algebraic setting. The formulation "solve the equation" may refer, according to different pupils, to various practices in the algebraic setting:

I *recognize* a degree 2 expression, I *know* how to solve $ax^2 + bx + c = 0$, I *will write* the given expression in this form and then I *will apply* my method.

Let us describe the *tools involved: distributivity* of multiplication with respect to the addition, *associativity* and *commutativity* of monomials so as to get an expression $ax^2 + bx + c = 0$, *discriminant* $b^2 - 4ac$, *expression of the roots* in terms of the coefficients *a, b, c.*

I *recognize* an algebraic expression with products. I *am used to developing this kind of expression.* I *develop.* But I *don't know* how to deal with the new expression and I cannot go further.

However, this formulation "solve the equation" can induce a research of meaning which takes into account the whole statement $x \rightarrow P(x) = x^2 - 4 + (x + 2)(2x - 5)$ is a function. The issue is to find for which values of x this function takes 0 as a value. Here again, according to the pupils, there are several possibilities in order to find the zeros of a function. Let us indicate 2 scenarios:

– *Scenario* 1 with pupils who have algebraic knowings.

I *recognize* a polynomial function. A part of *P(x)* is in developed form, a part is given as a product of factors. I *know* that a product vanishes as soon as one of the factors vanishes.

The procedure described below reflects the implementation of these knowings.

I *recognize* that $x^2 - 4 = (x + 2)(x - 2)$ is a remarkable identity.

I *identify* a common factor. I put $(x + 2)$ as a factor, and I *get* for *P(x)* a factorized expression $(x + 2)(3x - 7)$. I *make* each factor *vanish* successively, and I *obtain* the two solutions for the given equation.

Let us describe the *tools involved*: a *remarkable identity* $x^2 - 4 = (x + 2)(x - 2)$, a *factorization, associativity* and *commutativity* of the operations, *conditions* for a product to vanish, *method* to solve a degree 1 equation with one unknown.

– *Scenario* 2 with pupils for whom the adapted algebraic pieces of knowledge are not useable tools (perhaps they have been met only as cultural objects, perhaps they have not been met at all).

I *know* what calculation to make. I *could try* values of x to make $P(x) = 0$. But why choose one value rather than another, and if $P(x) \neq 0$ how to "improve" x.

In this context, other working settings than the algebraic one may be concerned. Other tools than those of the algebraic setting may be put to work in order to further the question.

In the *graphic setting*, the graphic representation of pairs (x, P(x)) obtained with a calculator is a *pertinent tool* for some pupils who are used to it. With such pupils, the following procedure may be observed:

- sufficiently many points are marked and linked so as to make an acceptable curve, with points on each part of the x axis. Then, the absciss x_A of a point A of the curve on this axis is read. The control is made in the numerical setting.
- the issue is to check if $P(x_A) = 0$. If yes, x is a solution of the equation. If not, one can improve the precision with successive estimates.

In the *numerical setting*, one tries to reduce the distance between two values x_1 and x_2 for which $P(x_1)$ and $P(x_2)$ have opposite signs, until one gets, if possible, one for which $P(x) = 0$.

Note that the numerical procedure cannot alone induce conjectures on the numbers of roots, while the graphic procedure may do so.

Let us mention some implicit tools involved in the procedure described above, the graphic one and the numerical one:

- in the graphic setting, the *graph* of the function $x \to P(x)$ is *connected*
- in the analytic setting, the *function* $x \to P(x)$ is *continuous*, and the *intermediate value theorem* asserts that, if $P(x_1) > 0$ and $P(x_2) < 0$, then there exists $x_3 \in [x_1, x_2]$ such that $P(x_3) = 0$.

2.3. *Tool and Object – Changing of Settings*

We say that a concept is a *tool* when the interest is focused on its use for solving a problem.

A tool is involved in a specific context, by somebody, at a given time.

A given tool may be *adapted* to several problems, several tools may be adapted to a given problem.

A tool can be implicit. It is the case for a concept being elaborated. It can occur under various forms: actions resulting from non-formulated intimate conviction, declarations without justification, possibly explicit implementation of techniques and familiar practices concerning an object not yet constructed as we saw in the examples above.

A tool can be explicit. This is the case when there is an intentional implementation of a piece of knowledge.

The *meaning* of a concept is derived from the context in which it is involved. Thus, it is the status as tool which is concerned. It is derived also from the relations developed in the context with other concepts in the same mathematical domain or not. The network of relations depends on the problem and also on the scientific and cultural attitude of the problem-solver. This attitude has as its elements, personal knowledge, habits and a field of references.

A concept is an *object* when it is considered in a cultural dimension, as a piece of knowledge independent of any context, of any person, which has a place in the body of the socially recognized scientific knowledge. An object is mathematically defined. This can be by various means: properties, an effective construction, or by an existence theorem.

The status as object allows the capitalization and the structuration of knowings and the extension of the body of knowledge.

The aim of the learning consists in the acquisition of knowledge in a double dimension: operative and cultural.

The *changing of settings* is a means to obtain different formulations of a problem. It can be a source of imbalance. That is the case if the formulations are not exactly equivalent, but partly corresponding and if the habits and the knowings of the researcher in each setting provoke an imbalance between his expectations and his observations or the mathematical results. The interest is in a new approach to the difficulties met and in the activation of tools and techniques not clearly relevant in the first formulation. We saw this phenomenon in the examples above.

One can read in the review *Recherches en Didactique des Mathématiques* (1983 p. 290) a study by G. Arsac about another historical example where he uses the notion of changing of settings in his analysis.

The study of G. Arsac turns on the irrationality of $\sqrt{2}$ in the arithmetical setting, and in the geometrical setting on the incommensurability of the diagonal of a square with the side.

The key-point is the following: the conviction, at that time, was that two magnitudes always share a common aliquote part, that the Euclidean algorithm can detect. However, this process goes on for ever when it is applied to the side of the square and its diagonal. On the other hand, the problem stated in the geometrical setting could immediately obtain an arithmetical formulation due to Pythagoras' theorem: "find a number a such that $a^2 = 2$". In the arithmetical setting, "the search for the square root of 2 may be difficult and lead to approximations without suggesting the idea that this square root does not exist ... In the geometrical setting, the Euclidean algorithm process leads to doubt about its existence. A return to the arithmetical setting can induce the hypothesis of non-existence of the square root, in the light of geometrical experience. Now, in the arithmetical setting the proof of the non-existence requires only very simple tools: the reasoning by contradiction and the classification of numbers as even and odd numbers. At least, the last one was familiar to Pythagoreans. Assuming the irrationality is established in the arithmetical setting, we get, if not an explanation, at least a result which makes the difficulties encountered with the diagonal of the square less incomprehensible. However the geometric setting remains a problem, until the solution is brought by Eudox's theory of ratios, presented in Euclid, chap. X".

2.4. Settings and Conceptual Windows

The notion of *setting* has at least 3 dimensions: a mathematical dimension, a socio-cultural one and an individual one.

In its *mathematical dimension*, a setting is constituted by objects in a given mathematical domain with various levels of complexity, by relations between these objects and by their various formulations. For instance, the numerical setting is constituted by numbers, numerical operations and the order relations, but also ordered pairs of numbers, numerical functions with all the properties necessary for computations ... Many other objects are to be included in such a setting, but it would not be possible to make an exhaustive list because this list has a cultural character. That is where the *socio-cultural dimension* comes in the picture. Persons with the same culture or the same background, embedded in the same environment, facing the same problem, will usually mobilize the same objects, or at least in the same categories (even if the same analysis leads them to opposite decisions). Persons with different cultures, but sharing a common part, will be able to communicate using their common background and enrich their exchanges with their diversity.

The *individual dimension* is involved insofar as the constitution of a culture is not only the fruit of a social education – including at school – but rather the fruit of an interaction between the social environment and the subject. Therefore in a society – and we include the school society – individuals sharing a wide common field of reference, build for themselves representations and personal mental images and life experiences, including cognitive experiences.

The objects of the algebraic setting, or of the graphical setting, or of the geometrical setting, and the ways of dealing with them are not identical for two different interpreters such as a teacher and a pupil, or two different pupils. As we see, the setting involves the subject and thus evolves with the subject. We express that by saying that the notion of setting is an open notion.

On the other hand, a problem which is rich and questioning for one who tries to solve it may be formulated in one setting. Its study will usually activate other settings and related concepts coming from different settings.

The individual dimension of the problem-solver will be involved in his analysis of data, of the questions and of the questioning which will lead to his solution. A given searcher may be led to change the setting during the study. Various searchers may tackle a given problem with different approaches, or place themselves in different settings. Communication will induce interactions between settings.

The notion of changing of settings is important here. It plays a productive role every time it creates an imbalance between what is expected and what is observed. Reequilibration is then a source of new knowledge. The scientific dimension comes through the mobilized objects which on the one hand are involved in relations in the problem, and on the other hand come from different settings.

We call *conceptual window* the set of objects and relations mobilized in a given problem, whatever the settings are from which they come. The questions arising during the study and the ways of dealing with them may lead to moving the window so as to focus it on tools which appear to be pertinent.

So, the notions of *setting* and *window* are complementary.

2.5. Our Epistemological Position

For us, *to do mathematics* is to solve problems, to ask questions, to adapt what we know to contexts which are not necessarily familiar, to make conjectures and to try to validate or reject them. It is to find, describe, exploit connections between questions which may, at first, seem to be quite foreign to one on the other.

To *know mathematics* is to be able to use it in different situations: where one is told what to do, or where it is possible to understand the statement of the problem, but where no method of solution is suggested. In these cases, we are chiefly concerned with mathematics as a tool.

To know mathematics is also to know how the notions are related one to the other from a cultural scientific viewpoint. In these cases, we are chiefly concerned with mathematics as an object.

Our epistemological position leads us to say that a pupil *has some mathematical knowledge* if he is able to implement its use as an explicit tool in problems he has to solve, whether there are clues in its formulation or not, and if he is able to adapt them when the usual conditions for use are not exactly fulfilled.

Concerning the learning by pupils, we consider that working with concepts as a tool calls for the global knowledge of the pupil, and tests their meaning and coordination. Thus this work is meaning-producing. Decontextualization allows the constitution of objects and the capitalization of knowledge. Work, where the tool and object aspects are involved in alternate and interactive ways, should allow the adaptation of concepts and their reinvestment in situations different from the ones which have produced them (school situations or use outside this framework).

3. DIDACTICAL CONSEQUENCES

3.1. Teacher's Work – Pupils' Work

The teachers acquire the knowledge which they will be required to teach generally out of context, as object. The subdivisions and relational organizations that they make will depend upon several factors, in particular upon the conception they have of the learning processes.

Let us describe 2 scenarios.

(a) The teacher presents the objects in a manner adapted to the content concerned *and* to his or her pupils.

- He can present definitions, examples, theorems with proofs if possible. In this case, the pupils must make sense of the ideas in order to use them as tools.
- He can show some prototypes and develop familiar uses. In this case, with respect to several constraints the teacher is inclined to reduce the content to algorithms. The pupils are thereby reduced to acquiring some practice in certain contexts.

In this scenario, *the pupils are faced with objects and must use them as tools.*

(b) The teacher may contextualize the content to be taught in various situations, and organize the development of an inter-setting dynamic in order to give meaning. Mathematics is involved here as tool and alter is decontextualised to take the status of object.

The work of the pupils is to understand the problem, to establish a scientific relationship in order to make conjectures, to examine the consequences, to control the result, to look for coherence, to confront the results with those of other pupils. It is *not* to read the problem as a pretext for guessing the teacher's expectations.

Here, *the pupils deal with tools, and must build them (or a part of them) as objects.*

Between the two opposing positions, in the long term, depending on the content to be taught and the pupils concerned, several other combinations are possible. And sometimes there is no choice.

A very important problem for the teacher is how to subdivide the knowledge into parts if he hopes to take initiatives and not to follow always the usual curriculum.

3.2. Subdivision of Knowledge into Parts

We want to characterize a distribution of knowledge which takes the meaning into account. This leads to stressing the tool aspect of concepts and organizing them, through adequate problems, into *conceptual windows* which may partly overlap.

Each window is complex: it consists of the knowledge activated in a given situation. It includes aspects of various concepts, situated in various settings. The network of relations in problems between several concepts contributes to the meaning of each one. In the example given below for a learning process, the window is made up of various plane surfaces, the area as a magnitude and numbers. In another situation, the window is made up of numbers, rectangles, magnitudes (lengths of sides, perimeter and area of rectangles), of variables and functions (to study the variation of the area of rectangles with a fixed perimeter), of graphic representations (acting as a symbolic representation of data, and also as a set of points producing conjectures through its geometrical study).

Windows are delimited in terms of the level of the learning. Clearly, for a given concept – say area of plane surfaces – the didactical choices and hence the conceptual windows will be different at the beginning of Secondary School and in University. The union of such windows covers the curriculum of a given period in school – and may actually extend beyond it. The excess occurs essentially when implicit or practical tools (Douady, 1984) are put into action, contributing to the constitution of the meaning of the mathematical concepts to be taught.

3.3. About Problems

Before proposing conditions about problems which can have a place in the learning process, we give an example:

Could you find a rectangle with fixed perimeter and area in two cases
(1) the semi-perimeter is 41 cm and the area 402 cm
(2) the semi-perimeter is 39 cm and the area 402 cm.

This problem can be solved by algebraic means. But pupils who have these algebraic tools at their disposal are not concerned. The problem is interesting in the learning process for young pupils (11–14 years old) who have some knowings about rectangles. They can recognize the shape, describe some properties of the sides, the perimeter and the area of a rectangle, where measure units are chosen, and the corresponding numerical relations.

Such pupils can understand the statement of this problem, can conceive an answer and recognize whether a certain rectangle is a good answer or not, even if they have no method of finding it.

To progress in the study, such pupils need some means. Here, with the knowings supposed, they can choose a rectangle satisfying one of the two conditions and control whether the other is more or less satisfied.

The problem which becomes interesting is that of learning how to deal with an approximation situation in the first question, and how to organize a proof in the second question.

In this problem the geometrical and the numerical settings interact explicitly. The numerical formulation of the problem is:

How to find two numbers a and b such that

1st case $a + b = 41$ and $a \times b = 402$ 2nd case $a + b = 39$ and $a \times b = 402$.

Thus, the problem is to solve a system of two equations with two unknowns. The pupils concerned do not have the algebraic knowings to do that easily. Their habits lead them to seek to transfer the problem into another setting where they have other means at their disposal. In this situation, their habits of numerical variables lead them to "read" the unknowns a and b as variables and the equations as two relations. Each one defines a set of pairs of number which can be represented graphically. Here a new setting seems pertinent to deal with the new formulation of the problem.

We can think of two scenarios.

(1) One graph to represent $a + b = 41$ followed by a numerical study to take into account the other relation. The graph allows them to choose new pairs of numbers better and better, using an implicit tool visualized by the graph: the function $a \to 41\text{-}a$ decreases.

(2) Two graphs, on the same grid, one for each function followed by a search for common points, a reading of the coordinates of these points if they exist, then a numerical determination of the graphical solutions.

By reference to the example above, we can follow the conditions formulated below.

Conditions for the Problems

The problems we chose as learning sources are submitted to the following conditions:

(1) The statement – context and questions – has a meaning for the pupils.

- The context refers to a familiar domain (within or out-of-school).
- The questions need answers which are not obvious for the pupils, but they can imagine what an answer could be, independent of their ability to propose a procedure for tackling or validating it.
- In view of their knowledge, the pupils are able to initiate a solving procedure, and to control its effects.

(2) The pupils cannot solve the problem completely, for reasons which may be diverse: the procedure considered would be too long, would cause errors or be incompatible with the time allowed; it would have to be used outside of its known field of validity, and possibly it does not work outside of its familiar field of functioning. There is a contradiction or a conflict between what the pupils conceive and the demands of the problem.

(3) The knowledge aimed at by the learning (content or method) is made by the tools adapted to the problem.

(4) The problem can be formulated in at least two different settings.

3.4. Our Proposition

We propose a model of the school relationship between the teaching and learning of a given piece of knowledge called *Tool-Object Dialectic*. This model supposes the existence of problems satysfying several conditions. This Dialectic is lined by an *Old-New* Dialectic, and it involves Interplays Between Settings which favour the development of coordination between concepts acting in different ranges. We shall describe this model.

Interplays Between Settings

By *Interplays Between Settings* we mean changes of settings brought about by the teacher, within suitable chosen problems.

The aim is to speed up the phases of the research, and specifically to induce a chain of relevant questions for the problem (which takes place in a definite situation).

It is also to prepare the *decontextualisation* of the tools engaged in the problem. In this process, three phases can be distinguished:

(1) *Transfer and interpretation*

Pupils are faced with a problem formulated in a certain setting. Taking into account their knowledge, their uses and habits, they are led by the analysis they make of the problem to translate all or parts of it into a different setting and interpret there some of the questions. In doing this, they activate correspondences between different settings (between objects and between relations).

(2) *Imperfect correspondences*

The correspondences between settings may however be only partial, either for mathematical reasons or because of insufficient knowledge of the pupils. The situation then becomes one of imbalance.

(3) *Improvement of correspondences and progresses of the knowledge*

New objects may be defined by extending correspondences, known relations may be extended and new relations created. The accomodation process is favoured by communication between settings, specifically by communication with an auxiliary setting of representation. This is a re-equilibation factor.

Tool-Object Dialectic

Given a suitably chosen problem, which the pupils are to solve, the *Tool-Object Dialectic* is a process in several phases. We describe these phases below and illustrate by an example (implemented in several classes, pupils 11–12).

(a) "Old, Explicit Tools". Pupils can mobilize objects of knowledge which they know, as *explicit tools*, to start a solving process or actually to solve a part of the problem.

Example: *you work in a communication situation. Each transmitter draws a triangle and sends a message (written without drawing) so that the receiver draws a superimposable triangle.*

Every pupil can draw a triangle and measure its sides. They all know how to control by transparent superposition whether two triangles coincide. All pupils can start.

Working in a situation of communication means that each pupil is a transmitter and a receiver to and from another. When a message is not understood by the receiver, he can ask a written question of the transmitter. The teacher acts as mailman.

(b) Research "the Implicit New". In view of the pupils' knowledge, it is unlikely that the problem will be solved completely, the first strategy being costly (in number of operations, risks of errors, uncertainty in the result...), or totally inefficient. *The aim of the teaching is the adapted tool* for solving the problem.

Each one (individual or group) knows that he is supposed to make proposals starting from what he knows, and that he will have to argue and confront his proposals to those of others in the eight of validation.

In the example above, to draw a triangle without any constraint is one thing, while to draw one satisfying given conditions is something else. Depending on the content of the messages, the task is more or less difficult.

The transmitter may send sufficient information, for instance the measures of the three sides. The receiver may exploit them correctly, here with the use of a compass. This construction is not to be greatly expected. One expects rather the drawing of a first segment and then locating the third point at the given distance from the first two, by approximation using the graduate ruler; this being usually unrelated with the use of the compass, known only in different circumstances (drawing friezes...). The receiver may not succeed, and ask for extra information. That is the contract.

The transmitter may send insufficient information such as the length of a horizontal side and an indication of the height of the third vertex. The receiver may provide a triangle satisfying the conditions of the message but not superimposable, because one piece of information is lacking. The attention is thus drawn towards what makes the difference between the two triangles, which can be expressed in terms of lengths or angles. It may be superimposable in spite of insufficient data, if both the transmitter and the receiver have drawn an isoceles triangle.

This is a phase of "action". However, conjectures can be put forward. This allows the *implicit use of new tools* Their novelty is in the extension or the change of their validity domain (as here with the compass used to select points, or the circle viewed as a set of points), or in their very nature (an aspect of the concept of an angle, of orthogonal projection of a point onto a line). Schematically, we can speak here of *implicit new.*

In these phases of action and formulation, efficient progress is often produced by a change of setting for the work: numerical formulation of a geometrical problem, graphical interpretation of the numerical form ... (cf. an example in R. Douady, 1985). This may be an opportunity to activate implicitly new tools.

(c) Making Explicit and Locally Institutionalized. Confrontation of productions and justifying statements. Some elements have had an important role in the previous phase and can be appropriated now by the pupils. They formulate these elements in terms of practice with their local use (here the measuring of angles), or the teacher formulates them in terms of objects.

In the example above, the point is to explain what affects whether a drawing will be reproduced correctly or not, and to state conditions which will guarantee reproducibility for any initial drawing. Here the measure of the three sides, or else the length of one side taken as a basis, the height of the third vertex and the position on the base of the projection of the vertex. Without this last datum, several triangles can be drawn which preserve the other data while letting the base angle vary. Thus there are other ways of selecting a triangle: for instance, by giving the length of one side and the measure of one of the two angles involving this side and the height of the third vertex. Here is an opportunity to institute the use of the protractor to measure angles.

As we said, tools are contextualised and personalised.

In this phase, the work and the proposals of the pupils are discussed collectively. In a "situation of communication", knowledge diffuses in various ways depending on the pupils. Each one is concerned in his/her own way by the different viewpoints. The negotiation between pupils or between the teacher and pupils facilitate the *depersonalisation* of tools without necessarily going out of the context. That is necessary to establish a common understanding with the class, an agreement about what will be the class knowledge, that each one can refer to. That is a function of the next phase.

(d) Institutionalization – Status of Object – Explicit New. Among the knowledge which has been made explicit in the previous phase, the teacher

selects a part to be decontextualised, depersonalised, a part that he/she will present with the usual conventions in a lecture and require it to be memorized. He "gives the course" presenting well organized definitions, theorems with proofs if necessary, emphasizing what is important in contrast to what is secondary, with respect to the mathematics and also to the pupils. This concerns knowledge and skill which are a part of the mathematical culture, which can be reused within the class or outside it. In doing this, the teacher provides them with the status of *object*. We speak here of *explicit new*, which later will be taken as old: here the role of the compass to draw segments of a given length or to select a point, the signification of the circle as a set of points. A theorem: a triangle is known, up to displacement and symmetry, when the measures of the 3 sides are known. Also possibly, the concept of height of a triangle from one vertex.

Giving the status of object to knowledge which has been only tools, up to then, the teacher constitutes *within the classroom a common reference* which enables each one to assess his/her knowledge, and thus to hope that their knowledge will progress.

On the other hand, personal structuration, which is significant in mathematics so that the knowledge is available, has been properly initiated in the former phases. However, in order to secure it, each pupil needs to test the knowledge he thinks he has acquired, and to sum up what he knows, all by himself, in different situations: repeated trials or new problems. A function of the two next phases is to develop and favour the transformation of collective knowledge into the knowledge of every one.

(e) Becoming Accustomed – Applying. The teacher asks the pupils to solve various exercises which require as explicit tools what has been recently institutionalised. In doing this, pupils may develop good habits and integrate the common social knowledge with their own particular knowledge.

In our example, the teacher gives several triples of numbers and asks for triangles having these numbers as side-lenghts. As an other task, he fixes the length of two sides, and let the pupils choose the third one. The question is to study the numerical relation between the measure of the angle between the two given sides and the measure of the height relative to one of these sides. Measures are actually taken, using the graduated ruler and the protractor.

(f=a) More Complex Situation or New Problem. The *new object* become familiar is taken *as old explicit tool* in a more complex situation or a new problem involving other concepts which may be known (but separately learned) or may need to be learned in a new cycle of the Tool-Object Dialectic.

To continue with our example:

(1) three numbers are drawn at random between 0 and 20. One wants to draw a triangle with these three numbers as measures of side lengths.

Is it always possible? If yes, give a construction for the triangle. If not, give a triple which cannot be associated with a triangle.

(2) (In a second stage, in high school) find a way to predict in advance for a given triple whether or not it can be associated with a triangle.

We describe below another example of a situation compatible with the Tool-Object Dialectic (PME 9, Vol. II, p. 35–30).

Examples:

(1) How to determine and calculate the area of various rectangles which have the same perimeter?

This is an interesting problem for pupils who know how to answer if the measures are integers, and who are convinced that every rectangle has an area, but who do not know what to do when measures are not integers.

(2) Partition: colouring a grid in 3 colours.

A point with coordinates (a,b) represents a rectangle R with sides of lengths a and b. The problem is to compare the measure of the area $A(R)$.

T.O.D. Phase a: 'The Explicit Old...'

Mathematical concepts are implemented as explicit tools to begin to solve the problem.

In the examples mentioned above, pupils may suggest several rectangles and calculate the area using only integers (it depends on the numerical values involved). In the second problem, they can pick a point on the grid, read the coordinates, a, b, multiply $a \times b$ compare to 24 and colour it.

T.O.D. Phase b: Research 'The Implicit New'

Difficulties are met in solving the problem completely.

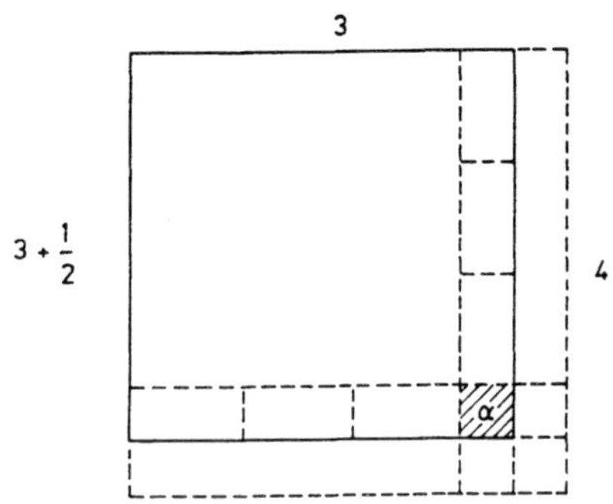

We come back to the first example above. If the perimeter is 16 cm, pupils must find two numbers a and b so that $a + b = 8$. Among these rectangles, they find the square (side of length 4 cm, area 16 cm^2). If the perimeter is 14 cm, pupils may find $(3 + 1/2) + (3 + 1/2) = 7$ and meet a difficulty in determining the area of the square with sides of length $(3 + 1/2)$ cm. In this situation, a phase of research begins for the pupils faced with this problem. They are led to modify their practice (which was to make a multiplication of integers) and adapt their knowledge. Here, they refer implicitly to the additivity of areas and thus reduce the problem to the following: How to find the area of the small square? Pupils are led to name the different pieces, to consider D as a part of the unit square and to exploit the fact that they can pave the square with four copies of D. Transferring this remark to a numerical formulation they obtain $4 \times a = 1$. The answer is $a = 1/4$.

More generally they are led to manipulate the situation to extend the correspondence length $\rightarrow$ area to non-integer measures. Numerically, they are led to extend the multiplication to rational numbers. The question is: how and with which meaning? Pupils will answer by making an interplay between the geometrical setting and the numerical one as we have suggested above. We will come back to this later.

In the colouring problem, the procedure described (in phase a) allows the colouring of a few points (10 or 20). However pupils must find an economic algorithm to colour all the grid (a thousand points or more). Actually they use the compatibility between order and multiplication to progress. This property is an implitic tool. It is implemented with integers, and it is extended to 'measures between integers' which pupils do not know how to designate. They declare:

- "A red point, more than 24, above a red point and on the right, even more, Red points".
- "Below a blue point and on the left, blue points".
- "Below a black point, blue points; above a black point, red points. Black points are more interesting than the others".

These are only the striking steps. Actually in the classes, they are separated by intense waiting periods which vary a lot with the pupils and with the numerical values involved.

Later, pupils will look for the 'square 24' (that means the square with area 24 cm^2) either "at the crossing between rectangles with area 24 cm^2 and the squares" or numerically: what is the number x so that $x \times x = 24$? Is it possible to find such an x?

The problem is how to choose, designate, calculate with numbers near the cross, closer and closer. This research leads to decimal fractions to approach such an x.

T.O.D. Phase c: Making Explicit and Locally Institutionalized

Some elements have had an important role in the previous phase and can be appropriated now by the pupils. They formulate these elements in terms of practice with their local use, or the teacher formulates them in terms of objects. We call these '*new explicit tools*' with which the pupils may become familiar.

Thus, the first problem above leads pupils to extend the multiplication to the fractions they know. For instance:

$$1/2 \times 1/2 = x \quad 4 \times x = 1 \quad x = 1/4$$
$$3/8 \times 5/8 = (3 \times 5)x \quad (8 \times 8)x = 1 \quad x = 1/64$$
$$3/8 \times 5/8 = 15 \times 1/64 = 15/64.$$

This computation results from an interplay with the geometrical setting.

The second problem, involving the three geometrical, numerical, graphical settings, leads to another problem: how to find the black points?

This problem induces the pupils to give a meaning to a/b (where a and b are numbers, integers or non-integers) through an interplay between the three settings above: that is, the x such that $b \times x = a$.

This moment is convenient for developing the process of division.

The research of the square with given area 24 cm^2 or 27 cm^2 ... involves computation and comparison with various numbers and forms of decimal fractions $P/10$, $P/100$, $P/1000$ ... . At that moment, an economical writing form is needed. Pupils may propose several codes adapted to this need.

T.O.D. Phase d: Institutionalization – Status of Object

The teacher presents what it is new and is to be retained, with the usual conventions. In our examples, that is the writing of decimal numbers, computation with them, comparison, their property to approach every measure as near as is wished.

T.O.D. Phase e: Becoming Accustomed – Applying

T.O.D. Phase f=a: Sophistication of the Task or New Problem

New knowledge is used in a more complex situation involving other concepts which may be known (but separately learned) or may need to be learned.

Example: Could you find a rectangle such that:
- The semi-perimeter is 41 cm and the area 402 cm^2.
- The semi-perimeter is 39 cm and the area 402 cm^2.

From now on, the new knowledge becomes '*old*' and a new T.O.D. cycle can begin.

CONCLUSION

We have presented some important elements for describing and analysing the relations between the teaching and learning of mathematics in school or university.

Thus, from the point of view of mathematics, we have distinguished

- the status "tool" from the status "object" for mathematical concepts.
- the settings in which they are located.

We have stressed the importance of the relations, both between concepts within a setting and, more importantly, between concepts of different settings. We have stressed also the significance of changing settings.

From the point of view of the participants in the didactical relationship – teachers and pupils – each arrives in class with a body of knowledge and a set of beliefs and customs, social and cultural, both conscious and unconscious. The relations which these participants will establish and develop with mathematics are conditioned by their metacognitive representations. These representations have a mathematical element as well as psychological and social elements. The didactical organisation – the Tool/Object Dialectic – which we propose is a model which takes into account these elements. But it is only a model with a certain domain of validity. The constraints of the class may make it necessary to combine the realisation of this model with other forms of didactical relations. The articulation of these different forms in the long term are the objects of study for the didactics of mathematics.

REFERENCES

Artigue, M. and Douady, R.: 1986, 'La didactique des mathématiques en France, émergence d'un champ scientifique', *Revue Française de Pédagogie* **76**, 69–88.

Artigue, M. and Robinet, J.: 1982, 'Conceptions du cercle chez des élèves de l'école élémentaire (7–9 ans)', *Recherches en Didactique des Mathématiques* **3.1**, 5–64.

Bachelard, G.: 1977, *La formation de l'esprit scientifique*, Vrin, Paris, 10e éd.

Bishop, A.J.: 1980, 'Classroom conditions for learning Mathematics', *Proceedings of PME* **4**, Berkeley, California.

Boero, P.: 1989, 'Mathematical Literacy for all, Experiences and Problems', *plenary lecture, Proceedings of PME* **13**, Paris, 52–76.

Brousseau, G.: 1986, Fondements et méthodes de la didactique des mathématiques', *Recherches en Didactique des Mathématiques* **7.2**, 33–115.

Douady, R. 1985, 'The Interplay between Different Settings; Tool-Object Dialectic, *Proceedings of PME* **9**, Noordwijkerhout, Vol. 2, 33–52.

Douady, R.: and Perrin, M.J.: 1989, 'Un processus d'apprentissage du concept d'aire de surface plane', *Educational Studies in Mathematics* **20**, 387–424.

Hoyles, C. and Noss, R.: 1988, 'Children Working in a Structure Logo Environment: From Doing to Understanding', *Recherches en Didactique des Mathématiques* **8**, **1/2** 131–174.

Joshua, M.A. and Joshua, S.: 1988/1989, 'Les fonctions didactiques de l'expérimental dans l'enseignement scientifique', *Recherches en Didactique des Mathématiques*,

première partie, **8.3**, 231–266, deuxième partie, **9.1**, 5–30.
Perret-Clermont, A.N.: 1979, *La construction de l'intelligence dans l'interaction sociale.* Peter Lang, Berne.
Piaget, J.: 1975, *L'équilibration des structures cognitives*, PUF, Paris.
Robert, A. and Tenaud, I.: 1989, 'Une expérience d'enseignement de la géométrie en Terminale C', (section scientifique, élèves de 17–18 ans), *Recherches en Didactique des Mathématiques* **9.1** 31–70.
Schoenfeld, A.: 1985, *Mathematical Problem Solving*, Academic Press, Orlando.

ENTR'ACTE 6

Where Douady's analysis concerns the cycle of development of mathematical knowledge through the tool/object metaphors, Werner now moves the reader into the living classroom inhabited by complex human beings. In particular he focuses the reader's attention on the breadth and richness of connections which children can make with mathematical ideas. His contribution complements that of van Dormolen's earlier, which looked at representations of mathematical knowledge through the eyes of the teacher. Here Werner makes the reader look at that knowledge from the child's position. His sub-text, unelaborated but all-pervasive, concerns the critical importance for teacher and educator of 'observation'. He cautions against over-theorising and encourages the teacher to stay at the level of observing the child's activity. His message is a warning that it is impossible for a teacher to know what lies underneath each child's behaviour. Just as mathematical knowledge is a fragile thing, so, of course, is interpersonal knowledge. On the other hand it is possible to develop a promising activity, and it is possible also to be surprised.

TAGE WERNER

OBSERVING CONCEPTUAL COMPLEXITY

1. OBSERVATION

In teaching/learning situations observations are inevitable. For the sake of teaching, the teacher at least must observe the pupil.

Teachers observe pupils, and pupils observe their classmates and their teachers. Researchers observe special traits chosen from the situations. In the teaching of mathematics many different things have been systematically scrutinized, and for many different purposes, and as the teaching of mathematics is a very complicated matter there always will be lots of things to concentrate upon. Mathematics is complicated – not least in its school-appearances, and children are very complicated – not least in their roles as pupils and students. They do not go to school with their heads only. They also bring their stomachs and their hearts.

In the first part of this century Harold H. Anderson in his observations of teaching situations in the USA focused upon the actions of the teacher. In the seventies John Elliott in Cambridge tried to interpret the ideas behind what the teachers meant they were doing and what an observer of the teaching situation, and the students in it, felt about it.

During many years a variety of observations has taken place under the name of "Error Analysis" and in areas closely connected with this, and different consequences have been drawn. Examples are provided by the work of R.B. Davis, M.A. Clements, E. Puchalska and Z. Semadini. Recently also some aspects from "Problem Solving" have caused certain observations of learning situations. After the great inspiration from heuristics in the forties, – above all caused by George Pólya – had faded for a period, factors such as the first recommendation in the "Agenda" from NCTM and the work by Alan Schoenfeld and many others have awakened new interests around these kinds of questions, now understood in new ways.

At the Royal Danish School of Educational Studies in-service courses are run for teachers in the Danish school. Here teachers also through further-education can obtain a master's degree, and they can be conferred a doctor's degree.

Some of the courses are intimately related to the research at the institution, and the following examples of observed reactions, remarks, imagery, use of metaphors, etc. among learners of mathematics, have all been collected and discussed in connection with these courses. The teachers have read about

Alan J. Bishop et al. (eds.), Mathematical Knowledge: Its Growth Through Teaching, 133–143.

observations of students in mathematical situations – for instance texts from the above mentioned researchers – and they have reported and discussed examples from their own experiences. As the institution provides no initial education of teachers, but solely further education, all the students are mature, experienced teachers, and their personal learning processes take place in a context in between the contents of the courses they attend and their daily work in the schools. Thus the examples given in this chapter will try to add some other colours to the above mentioned kinds of "observations".

In the Danish schools all the children aged 7–16 work 4 lessons every week with mathematics. Behind the official decisions about this lie different notions of what the expected impact of going to school should be. At least it must include these three: First, the manysided development of every single child as a human being. Second, the ideas of bringing up and of the socialization of the child as a member of certain groups: The family, the village, etc. Third, the acquisition of knowledge and skills.

Sometimes a teacher in his observations of a pupil is solely interested in the last mentioned results of teaching – and that may of course be very respectable. But sometimes a teacher forgets about one, or two, of the three impacts, and that might lead him to misinterpret the situation. Thus – and in other ways – there are obstacles for a teacher who honestly tries to do his best for his pupils and to watch them fairly for the sake of "good teaching".

In Denmark the Act for the schools shows the guidelines for the teaching. For instance, it states what subjects must be treated, and with what weight. The Act does not – however – in any detail point out a curriculum. In connection with the legal situation only some main topics are mentioned as expected content in the teaching. For every school a more detailed plan is worked out by the parents and the teachers.

Of course all this complicates the situation for the teachers who not only are obliged to observe whether the pupils learn and behave, but also to observe how they develop, and how all this fits into the ideas and understanding of the pupils, the parents, the colleagues, and the authorities.

The fact that the situations in the classroom are always complex should not make a young, beginning teacher nervous and restricted, for example in such a way that he only can reduce his observations to questions like: Does the pupil do his sums correctly? On the other hand the richness of the situations should raise his curiosity and make him appreciate the many possibilities which it opens up. But of course it is not at all simple for a beginning teacher to handle all this, he may need some help from experienced colleagues. And maybe it may also be difficult to see many of the complexities, for example if one was brought up in a teaching-tradition where the teacher gave all the questions heard in the room, and the children were expected to answer in his (the teacher's) special style.

The following sections contain some attempts to point out where some of the complexities may be. And between the lines should also be read something about the fruitfulness of complexity and of relations between theory and praxis in teaching of mathematics. The "difficulties" mentioned should be understood as "nice ones" and not "nasty ones" – exactly like "problems" in mathematics are good friends. Do not solve them too quickly, because then you lose them and have to go and find some new ones to fill the gap.

2. ON "CONCEPTUAL RICHNESS"

In mathematics lessons it sometimes happens that a pupil realizes something which for him at the moment seems to be quite new and unknown – although it deals with things that he thought that he knew exceedingly well. Experiences of this sort may show the pupil that well-known matters often can reveal unexpected aspects, and that you can get very well acquainted with mathematical ideas without obtaining complete knowledge or understanding. This richness and complexity in potentiality for thinking and feeling in mathematical situations puts some demands upon the teacher, both when he addresses a pupil, and also when by listening he tries to get an impression of the pupils' ways of understanding.

Let us try to illustrate by an example how objects from the realm of mathematics – objects which might at a first glance appear innocent, modest, and uncomplicated gadgets – in the mind of a child can have a richer and more colourful nature.

Now, for the sake of simplicity – let us choose the number eight as an example. The children will meet eight in grade one, and many of them will have met eight long before they go to school. What might grow up around this number in a child during the years in school? And outside?

Eight is a number – and a numeral – and as a natural number it can handle both ordinal and cardinal (and other) matters. It is seen in lots of sums in an arithmetic book. Eight is the number of tyres of two cars and of teeth in a certain jawbone. The width of a chessboard and the number of white pawns. For a child of seven it is something to look forward to. It has something to do with a major scale and with the keyboard of a piano. In some countries 8 is the same as 20 – in the evening. Funnily enough October is not the eighth month. Even if spiders are small, crawling creatures, they are not considered to be insects – because of eight. Some cuttlefish have eight arms.

Eight is an even number, and it is a cube. It is a sum of two squares, and also the difference between two squares. It is a close friend of 45 and 360, which is seen in both geometry and computations of interest. To say nothing of its relations to 12 1/2.

The presence of the number eight has some consequences. If you are eight in a group some games are nice to play, and others are impossible.

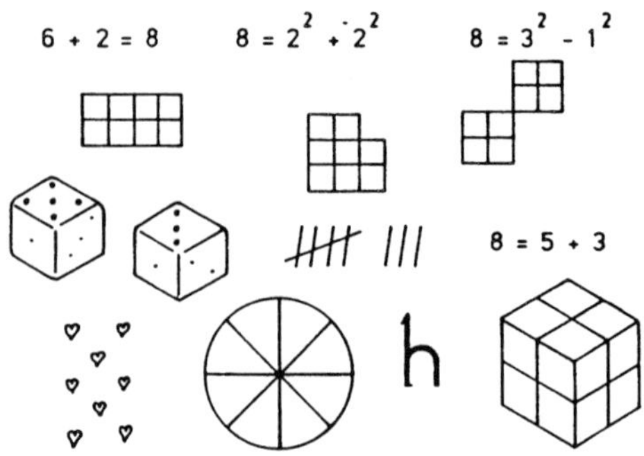

The fact that "five plus three are eight" may for an adult be just one question which should be learnt in less than ten minutes. But for a sensitive child it contains many – and maybe rather different – aspects. "Step 3 to the right", "give me 3 more", and many other expressions can be heard in school, and elsewhere.

Five plus three are eight.
Five apples and three apples are eight apples.
Five fingers and three fingers are eight fingers.
5 km and 3 km are 8 km.
5 cm + 3 cm = 8 cm.
5 km/h + 3 km/h = 8 km/h.
5 gnats + 3 elephants = 8 animals.
5 km plus 3 cm might be 8 something.

In a cottage in the dark forest a girl lays the table – waiting for the men to come home from their work in the mountain. She is as white as snow and as red as blood, and her hair is as black as the wing of the raven. How many plates does the girl put on the table?

Even if the situation is a mathematics lesson in which an adult accepts the role of a teacher and a child the role of a pupil, and very few of the facets of "number eight" are relevant, the child still is a child, and the teacher is co-responsible for the decisions of what should be regarded as relevant in the situations. Furthermore the situation itself is partly defined by what is accepted in it, and thus there are close bonds between the situations of teaching/learning and traits of knowledge and actions in the situations.

3. ODD NUMBERS

The ways in which the pupils represent some of the mathematical ideas, and the actions they perform on them, provoke them to use words which often reveal something about what is going on in the mixture of representations, images, feelings, and abstracted conceptual ideas in their minds.

This section contains some examples of what children have said in situations where parity more or less played a role.

In some classes the pupils represented numbers by wooden cubes. For instance, they built different "houses" each containing 5 "flats" (=cubes), and they showed that a town consisting of one 2-house and three 6-houses contained 20 flats, like a town consisting of five 4-houses. Among many other things some of the pupils sorted out 1, 3, 5, 7 to one side and 2, 4, 6, 8 to another. A girl of eight years was asked why she placed 7 exactly there, and she replied: "Because it is one of those with wings of equal length" – at the same time trying to explain herself better by waving her arms.

One pupil referred to 7 and to 11 as "every-second-numbers". When she was asked if 6 also is an "every-second-number" she hesitated a moment, then she said: "Oh yes. But one of the others".

In several classes odd numbers were mentioned as "autobus-numbers". In the situation the language of the children is not mysterious, but easy to understand. The "autobus-number" is not the number of a busline, but the expression comes from the shape of the representation:

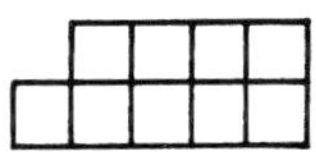

But even if this may appear easy to understand in the situation, the effects of the children's use of words and illustrations, which they partly choose themselves, is difficult or even impossible to grasp. Will the girl with the wings of equal length more easily realize that the sum of the first n odd numbers is n^2?

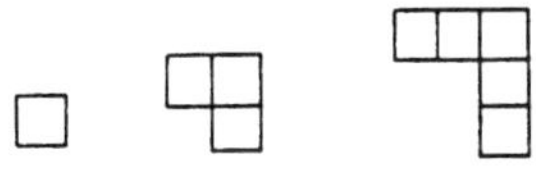

Do "bus-number"-experiences facilitate understanding of things like "odd plus odd is even"? But questions about the pupil as decisionmakers in the school situations are more profound.

Some 14–16 years old students, who were rather mature chessplayers, used the words “odd” and “even” precisely and frequently, when they were asked to explain details from chess – for example about kings taking opponents, or solving small exercises like

> White: *K* e1
> Black: *K h1, Kn a7, P h2*
> White in the move draws.

In situations where this picture was used – for example with questions like: Draw a red line from point *A* to point *B*. How often does the red line intersect the black line? Is it possible to draw a line, that intersects the black line exactly 3 times? 1 time? 4 times?

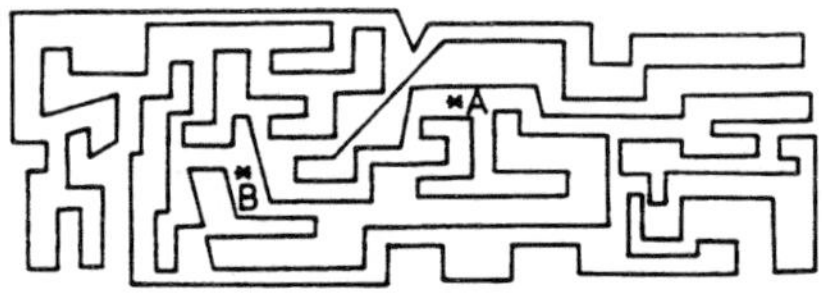

Younger children after a rather short time began to explain using terms like “inside” and “outside”, and they started arguing with words like this: “A is on the island, and B is out in the water, and therefore...”

It turned out to be much more difficult to make children explain by use of parity, why it is possible on a piece of paper to draw certain pictures without drawing a line more than once or lifting the pencil, while other pictures are easy to draw. And younger children were never heard to give any arguments in connection with this two-person “game”:

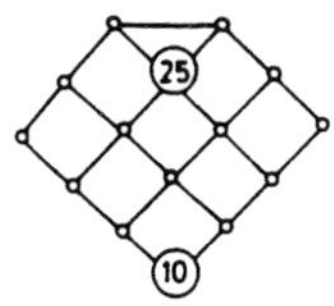

One player moves the 25-coin from a lattice-point to a neighbouring point – and alternately the opponent moves the 10-coin.

If the 25-coin can move to the place where the 10-coin lies at that moment, then he captures the 10-coin. The starting position is shown above, and 25 starts the game.

If 25 can capture 10 within his first 6 moves, he wins. If not, 10 is the winner. (10 cannot capture 25).

Most children say when they have tried the game some times, that it is very dull, and that the same player wins every time (which is correct), namely 10 (which is not correct, 25 is the sure winner).

4. CUBES

In some of the cases where children were observed while they worked upon mathematical questions, elementary shapes such as circles, squares, pyramids, and cubes were involved, for instance as mentioned in the above Section 3, or making "islands" from squares of equal size as the pupils did in the Nuffield project. Show all possible 6-islands and decide which might be cut out and folded to make a cube.

Or "building houses" – face against face – from, for instance, 3 red and 2 white cubes. Describe all 1-storied houses. Here is a 2-storied house – its roof is: 1 white and 2 red. Its outerwalls are: 5 white and 9 red. – Build it (or them). – The children themselves easily make 20 more questions, and work with them, and the teachers get opportunities to hear many interesting discussions. For example: It is not at all easy for a nine year old to count the walls of the house. A boy said: To the east there are 3 red and 2 white. To the north there are 2 red. And so on. Soon most of the pupils used this way of counting.

The cube is a very complicated thing to become acquainted with, but some of the pupils have seen that the octahedron and the cube are dual to one-another, that the tetrahedron lives in the cube, and the cube in the dodecahedron etc., etc., and for those students a feeling of understanding the cube is emerging. For some it seems to give a special sensation to see a cube via six pyramids turned over into a rhombic dodecahedron, or to see how a cube can be split into three congruent pyramids.

If the cubes are furnished with 1–6 eyes in the usual way, dice also can give some support to the cube with its six faces, eight corners, twelve edges, four long diagonals etc.

A girl of 9 years sat gluing eyes on cubes. She knew, that the sum of opposite faces should be 7. An adult visitor in the classroom took one of her dice, held it in front of the girl, and asked:

"Here you see a corner, and around it you see the faces one, two and three. – Do you think the die has a corner, around which you see the even numbers: two, four and six?" The girl answered: "I don't think so, and I don't believe so."

Now the visitor put the die away and asked the girl some questions, which the girl promptly answered. "Imagine I have the die here in my hand. With an even number against the ceiling. What would you know about the number against the floor?" "It must be odd." "Yes, and how many even numbers have we now left for the four walls here?" "Two." – "Very fine. We have got two even and two odd left for the four walls here." The adult was sawing with his hands in the empty air.

"Can the two even numbers sit opposite like this?" – His hands were still sawing up and down. – "No, they must sit like this." And now the girl also used her hands. "Now I ask you again: Do you think the die has a corner, around which you see the even numbers two, four, and six?" This time the girl answered: "Yes, here it is." And she pointed to a place in the empty air.

When the visitor proposed they should take a glance at the die she just made of cardboard, she was really morose and said: "No, we need not look, because now we know."

At a school four boys – considered by their teacher to be very good at mathematics – were presented by a visitor the question: "Here are three plastic cups – two with their openings and one with its bottom up. Turn two cups upside down at a time, and try with a series of turns to obtain the position, where all three cups stand with their openings up."

Now the teacher and the visitor left the boys alone. Half an hour later one of the boys came to the masters' room and said that they had solved the problem, and they went to the classroom to see what the boys had found out.

One boy said: "It is impossible. Because the position you gave us is in the green tetrahedron of the cube, and the other position is in the red tetrahedron."

The visitor was somewhat puzzled by this kind of reply, but the teacher of course fully understood the boy's message. And a "normal" Danish teacher would have been absolutely lost. Here are some cups, and the boy is talking about a cube? But the boys and all their paper on the table gave a good explanation. After some efforts about the problem: How to remember a series of turns? they invented this notation

010

Later this drawing was made

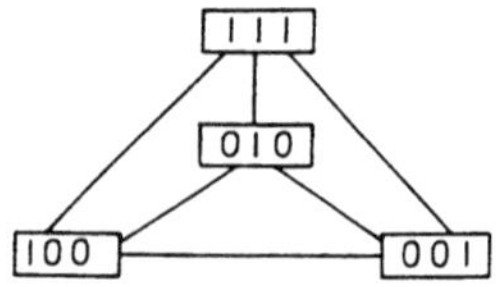

and now one for the boys saw, that this is a tetrahedron. And not only that, but it wore coordinates. Exactly like the cube, made of yellow sipping straws and furnished with a red and a green tetrahedron formed by the diagonals in the faces of the cube, they worked with half a year ago.

At that moment all of the four boys felt that they "understood" the problem, and that they were able to explain the result to the stranger.

5. OTHER DIFFICULTIES

If a student makes errors, for example in doing arithmetic, the reason may be that he uses some wrong strategies, and it may be difficult for the teacher to obtain a clear picture of what is going on in the child's head. But there are other reasons which are even more difficult to observe – sometimes they even seem to be impossible to observe, and we only dare guess.

The last two sections gave some examples where the teacher might have difficulties in understanding a situation, because of the pupil's use of mathematical concepts or some more private representations. In this section we will mention examples of difficulties more connected with personal traits.

A boy of 11 was considered by his teacher to be slow in arithmetic. A close look at the boy showed the curious fact, that the number 7 played a certain role for the boy. He managed exercises in which seven did *not* appear much faster and with less errors than the exercises containing 7. Nobody could give a good explanation.

Sometimes a teacher cannot avoid misinterpreting a given situation because of differences in the cultural backgrounds of the teacher and of the pupil. Take as an example a female teacher of 28, Maria, pleasant and intelligent, and trying hard to be a fair and effective teacher of mathematics. She lives together with her father, who is a dentist, in a very big house near the seaside.

Martin, 10 years old, is one of Maria's pupils, and he really likes her and appreciates her fairness. One day in a lesson Maria asks Martin: "What is 3 times 9?" Now Martin knows quite well, that the expected answer is 27, but he also knows, that his two close friends down here in the corner of the classroom do not kow this answer. And who is Maria? From his home he knows very well, that there are different kinds of people. His fther is a worker, and from his discussions with colleagues Martin knows about owners of factories and of office staff, and he knows that workers stick together in solidarity, and that the owners never should know all about "us" or what "we" can do. And who is Maria? Not a pal – rather a person from the other side of the school system.

When Martin does not answer her question Maria can only see one explanation: Because he does not know. He is a very nice and polite boy,

Martin, but he is a little peculiar. Sometimes he is very clever, but on other occasions he is quite different.

In cases like Maria's it often seems of great help that colleagues discuss their experiences with one another, and sometimes reading about other people's findings and about the principles or even theories grown up from the findings may help. Yet transformations by the teacher are always necessary.

Harold H. Anderson wrote down what he saw teachers do, and he tried to sort out the actions in three kinds: the good ones, the bad ones, and those in the middle. (To tell about his value-system is a longer story which should not be told here). The most serious blunder is (says Harold H. Anderson) that the teacher "Determines a detail of activity or acts for the child in carrying out a detail".

But even if a teacher understands and accepts this as a principle, it is not certain that he is able to stick to it in a given situation.

It is rather obvious that a teaching/learning situation in mathematics is very complex, both regarding the subject matter and the people involved. And this is very fertile – at least in the hands of a talented teacher. He can read theories about it and combine these with his own observations, and thus develop as a teacher. Yet, there may appear situations which he should not try too hard to "explain", but merely to accept.

As an attempt to illustrate how observations are not only banal necessities for the formation of private, personal principles to guide a teacher's actions in the classroom situations, but that some experiences may sometimes raise profound questions in a teacher, I shall conclude by briefly narrating an episode, which happened to me in 1955.

6. THE BOY AND THE PLASTIC MONEY

One can learn many things from reading books, of course. In the early fifties I read a lot of books – among other topics on mathematics, on teaching, and on psychology – when a boy in his own way taught me that living people around me also are sources of understanding, and that the observation of students in work with mathematics can support my private "theory of education" as well as reading Wynne's book "Theories of education" (which I find is a very good book).

The boy "hit" me so hard, that I shall never forget. Maybe because I was unarmed, and had no theory ready at hand to explain what I saw. The boy was a less-able 16 years old student (IQ about 70) at a school, where I studied the teaching of disadvantaged children. I saw him together with a teacher, who tried to teach him about money. They had a box filled with plastic coins: 5-øre, 10-øre, 25-øre and 1 krone (which is 100 øre). One of the problems for the boy was to change 1 krone into four 25-øre. He started by offering her three. A quarter later he was able to give her four every time. The boy was very bad in handling the coins, but he learned small bits, and he and the teacher clearly enjoyed the situation, they smiled and made jokes.

But funnily enough, day after day the same scene was repeated. They said nearly exactly the same sentences to one another, and the boy started with offering three 25-øre for 1 krone.

On the other side of the road there was a grocer's shop. The students of the school were not allowed to go there alone during the school-day. One day, as I went over there, the boy accompanied me. He had his own purse with money, and he wanted to buy some candy.

When he had received the candy I could not help it, but I tried to take the opportunity and said while his return money lay on the desk: "Oh, is that correct?!". And now he did it. He said without hesitation: "Yes, it is. I was to pay 1.35, and here are 65 øre in return".

ENTR'ACTE 7

In the previous chapter Werner points to the rich complexity of pupils' understandings of mathematics, and the difficulty for a teacher of knowing in any explicit detail what underlies pupils' overt behaviour. Nonetheless his chapter suggests that it is important for teachers to try to treat pupils on an individual basis and to take into account their thoughts and feelings. Werner ends by citing an example of the gap between the low performance of a boy in mathematics in school and his confident use of similar mathematics outside school using money in the candy store. This issue links to the next chapter of Hoyles, who, in attempting to draw out the particular features of informal computer-based learning environments, first seeks to characterise informal learning environments for mathematics in school and outside school. The contrasts between these environments are identified and some implications for teaching suggested. This leads to a description of informal computer-based learning environments for mathematics called 'microworlds', in terms which characteristise the software that forms the basis of the environments. The potential of microworlds in assisting pupils to discriminate the mathematical structures within their activities and also to generalize their understanding of these fragments of mathematics outside their immediate context of use, stretching towards the development of mathematics as a coherent system, is discussed. By this means an attempt is made to identify a teaching role for the software environments which is qualitatively different from explicit teaching and which complements rather appropriates the activity of the teacher.

CELIA HOYLES

DEVELOPING MATHEMATICAL KNOWLEDGE THROUGH MICROWORLDS

This chapter sets out to examine some aspects, both positive and negative, of the role of the computer in the learning of mathematics. The computer is not a unitary object. Its effects depend on the user interface and the software. Software exists along a continuum – from that which dictates the instructional sequence, (for example drill and practice programmes), through that in which the user may choose from a well-defined range of options for action, (for example simulations), to that where decisions as to action are not completely predetermined but constructed by the user (for example programming environments). These types of software stem from different educational paradigms and have very different positions within the pedagogic relations of a school class. They thus require different theoretical analyses as to their potential influences on learning. Throughout this chapter my concern will be solely with the third type of environment mentioned above; namely environments built with interactive software which is user rather than menu-driven. Within these environments, feedback is informative rather than evaluative and pupils have the opportunity to learn both from their actions on the computer and by reflection on the feedback provided. Pupils using this software – if allowed by the teacher – develop their mathematical knowledge through a process of inductive generalisation, through conjecturing and testing out their conjectures. The constraints under which they work and which define the mathematical nature of the activity are implicit rather than explicit. I will term this type of learning environment an informal learning environment which when computer-based is called a microworld. To provide a background for analysis of microworlds, I first gave a brief review of informal non-computer-based learning environments in mathematics.

INFORMAL MATHEMATICAL LEARNING ENVIRONMENTS

There is an increasing volume of evidence which shows how mathematics is used in everyday activities, for example in weaving, building, design, knitting, packaging (see Carraher *et al.*, 1985; Gerdes, 1988; Harris, 1987; Mellin-Olsen, 1987). Within these activities the mathematics is used to solve problems and to achieve a specific product or outcome. Thus the mathematics involved is a powerful means of "getting something done" (Douady's

Alan J. Bishop et al. (eds.), Mathematical Knowledge: Its Growth Through Teaching, 147–172.

work in this volume is of relevance here) and there is no explicit sense of the mathematical concepts embedded within the activity neither of the mathematical processes engaged in during the activity. The significance of the activity is not constituted in its different elements – here including mathematical elements – but in the dynamics of the activity as a whole. Thus the mathematical ideas are *necessarily* wrapped around by redundancy and mathematical understandings are developed as part of the relationships within these specific experiences and practices and not as separated entities. Additionally, these mathematical understandings are not synthesised into a global, rational and structured system.

Given the pervasiveness of mathematics embedded in everyday activities and the functional motivation which it stimulates – that is to produce a product – a crucial issue for mathematics education in schools is whether to incorporate or to ignore these 'real world' experiences. First the following question must be considered: How far can procedural knowledge, learning to do something, assist in the acquisition of conceptual knowledge, that is, a grasp of the theorems which structure one's actions *when viewed through a mathematical lens*? This has been the subject of some investigation by Carraher, Schliemann and Carraher (1988) in their studies of mathematical knowledge used in everyday life and in particular within the practices of fishermen and foremen. They found that 'both conceptual and procedural knowledge may result from practice with solving formal operational problems in everyday life' (p. 21). The fishermen, for example were able both to invert the computational strategies they normally used at work and to transfer them to other domains within which they were not usually applied. This provides some evidence of the presence of understandings *beyond* the solution of the immediate problem at hand and stretching towards an appreciation of more general mathematical relationships.

Many studies however have consistently shown that this sort of mathematics learnt informally outside school (abbreviated here to Informal Non-School Mathematics, I.NS.M.), even if extended beyond the mere application of procedures, forms part of the discourse of the activity and not the discourse of school mathematics; by school mathematics (abbreviated to S.M.) I mean the mathematical topics which form part of a school curriculum. Lave (1988a), for example, suggests that "Evidence and argument here support a claim that opportunities, such as routine meal preparation, which enable the realisation of measuring activity *don't* determine its structure, and may not share it. Structure results from and unfolds in the articulation of activities, social relations, and their settings" (my emphasis p. 130). Thus I.NS.M. and S.M. have different aims and different representations, and call up different strategies in their users.[1] Moreover, there is a further distinction between these two domains concerning who controls the goals or agenda of the activities: within I.NS.M. control rests with the users; within I.S.M. this, very evidently, is not the case.

This is a fundamental problem. Mathematics educators assert that their subject is useful and would like to think that it can connect with the 'everyday' intuitions of their pupils in order to give mathematics meaning and life. But, given the extent of the differences between I.NS.M. and S.M., mathematics educators must consider if these aims are feasible. The question is: *Is it possible to capture the power and motivation of informal non-school learning environments for use as a basis for school mathematics?*

One path that has been proposed is to set up informal learning environments for mathematics *in school* (abbreviated to I.S.M.) through projects and 'real' situations to investigate rather than "problems to solve" (see for example Papert, 1980; Mellin-Olsen, 1987). If these situations catch the pupils' imagination, they can be tackled creatively and relatively independently of the teacher. (Brousseau, (1988) would describe such situations as a-didactic.) Within I.S.M. it is intended that pupils come to use mathematics in a meaningful way within a context in which it is seen by them to be required. Pupils face the task of formulating their own problems, guided on the one hand by their own general goals and on the other by the difficulties they come up against during their activity. Thus, in a similar way to non-school activity, pupils working within I.S.M. derive their goals and strategies by a process of negotiation between their problem solving intentions and the constraints of the settings.

There still remain however fundamental differences between I.S.M. and I.NS.M. in terms of motivation and intention. The rationale for I.S.M. is to provide vehicles for generalisation and the formalisation of these generalisations. This contrasts with I.NS.M. where generalisations usually do not make sense since the activity is specifically geared towards a particular product. Additionally in I.NS.M. people frequently learn by observation of more expert practitioners through apprenticeships, a type of learning which rarely takes place in school. This is inevitable given the objectives of the school curriculum. The goal for a teacher is not merely the reproduction by the pupils of her actions (as in I.NS.M.) but the understanding of the principles governing her actions – a very different thing! In the past the significance of the role of the teacher in "abstracting" the mathematics from informal learning environments tended to be undervalued. Studies (for example Walkerdine, 1982) have shown that pupils do not discover mathematics simply by manipulating materials. Analysis of classroom interactions shows clearly that it is pedagogic intervention which imposes the mathematical structuring and provokes the pupils' awareness of the underlying mathematical ideas.

Table 1 below compares informal and formal learning environments in the two contexts, school and non-school. It points up the similarities and differences between I.S.M. and I.NS.M. together with the similarities and differences between I.S.M. and formal school mathematics (abbreviated to F.S.M.). By F.S.M. I mean the explicit teaching of mathematics from the

Table 1. A Comparison between Informal Learning Environments for Mathematics within School (I.S.M.) and Outside School (I.NS.M.) and Formal Learning Environments for Mathematics within School (F.S.M.)

Context: NON-SCHOOL (NS) | SCHOOL (S)

Type of Learning Environment: INFORMAL (I) | FORMAL (F)

INFORMAL (I) – NON-SCHOOL (NS)

Tasks are particularistic and generalisation does not make sense.
Apprenticeship modes of learning are often practised.
Mathematical content is not made explicit.
Group performance is rewarded; the product or the interpretation of the product is the vehicle of evaluation
Goals are set by and in the control of the users.

INFORMAL (I) – SCHOOL (S)

The rationale for tasks is to provide vehicles for generalisations and their formalisation.
Apprenticeship modes of learning are not practised.

INFORMAL (I) – spanning NON-SCHOOL (NS) and SCHOOL (S)

The mathematics is embedded in a task or a project.
The objects of activity are concrete (or visualisable).
Data is ill-defined and necessary information has often to be sorted out to provide data for the problem.
Precision and accuracy is defined by the situation
Numbers are "messy" and do not work out well.
Work is usually collaborative .
Correctness is negotiable and depends on the audience or the receiver of the product.
Language is imprecise and interpretable by reference to the setting.

SCHOOL (S) – spanning INFORMAL (I) and FORMAL (F) (dashed box)

Mathematical content is made explicit, named and the object of study through teacher intervention.
Individual performance is rewarded even if carried out in a group.
Goals are not set by and not in control of the users.

FORMAL (F) – NON-SCHOOL (NS)

[Hatched cell]

FORMAL (F) – SCHOOL (S)

Mathematics is decontextualised or isolated within a school discourse.
The objects of activity are abstract and/or symbolic.
Data is clean, well-defined and presented tidily.
Precision and accuracy is assumed or given.
Numbers are clean and work out well.
Answers are right or wrong; correctness defined by the rules of the discourse.
Language is precise and carefully differentiated.

starting point of its structures and relationships and allowable manipulative rules. I do not intend in this chapter to discuss in detail F.S.M. since my focus is informal learning environments. F.S.M. is included in Table 1 for completeness and to serve a comparative function.

With reference to Table 1, characteristics in the solid line box represent features of an informal learning environment whether school or non-school. Characteristics in the dotted line box represent features of school mathematics irrespective of the type of learning environment. Thus, Table 1 gives an overall summary but also serves to highlight the major differences between I.S.M. and I.NS.M., namely in the extent of the explicit characterisation of the mathematics and the teacher's role in this process.

In an attempt to identify the major features of I.S.M., we have devised and begun to refine a model which involves the dynamically related components of using, discriminating, generalising and synthesising (*U, D, G,* and *S*) (Hoyles, 1986). The essence of the model is to create situations in which:

- mathematical ideas and processes are used to produce an outcome (where there might be little appreciation of the nature of the mathematical tools being used);
- it is meaningful to break down the procedures used and discriminate the function of the components;
- the rewriting in explicit form of what was previously implicit is provoked and the appropriate tools to achieve this provided; or put another way the construction of generalised relationships is necessary for successful task completion;
- the evolution of the same mathematical structures within interrelated domains is encouraged in order to promote a synthesis across different contexts.

From the analysis in Table 1 it would appear that within non-school environments, I.NS.M., only the using component, U, tends to be present whilst the components D, G and S are generally activated in school by the teacher. In this chapter I wish to explore the following question: *How far within informal computer-based environments can the transition between the three components U, D and G be provoked through pupil/computer interaction without explicit teacher direction?*

What I am interested in here is not simply to add an intelligent tutoring system onto the informal learning environment but to identify qualitatively different ways by which the environment can provide support for learners in their search to understand and generalise mathematical ideas.

The fourth component, *S*, of our model, relating to synthesis is not the main subject of this chapter. However it is important to mention two points. First, pupils do not automatically use knowledge constructed within one domain in another and understanding tends to be restricted to the context in which the knowledge was first appropriated (Lawler, 1985; Hoyles and Sutherland, 1989). Connections between activities in computer and non-computer contexts are not necessarily made spontaneously, particularly if there are different representations of the mathematics in the two settings. Second, there remains the dilemma as to how the mathematical fragments drawn out locally in a variety of situations can ultimately be synthesised into a coherent system, a mathematical system, with its own rules and processes.

I now wish to speculate on the design of computer-based environments for mathematics; first as part of project work in schools with the aim that mathematics can be used in a genuinely problem solving way and second in order to build bridges between the discourse of the project activity and the discourse of a formal, mathematical system, following the components of the UDG model.

PROBLEM SOLVING IN A COMPUTER ENVIRONMENT

The computer with appropriate software has the potential to generate activity in a genuinely problem-orientated fashion in which pupils can experiment with ideas without necessarily any explicit theory, and can conjecture and test out hypotheses. The computer evaluates the pupil's actions but does not judge the pupil. Errors can therefore be treated as "bugs" from which to learn. Errors also can fruitfully be used as starting points for further investigations. The extent of independence pupils are allowed within this sort of activity is obviously a matter of negotiation between teacher and class.[2] However, with the presence of computers, the teacher has the opportunity to devolve more control to the pupils if they so wish since the pupils are able to look to computer feedback rather than the teacher to validate their responses. Additionally the pupils can benefit from knowing that the computer will not cheat or pretend; so their task is focussed on solving the problem at hand, rather than trying to 'guess what the teacher has as an agenda'. Thus a computer environment through interaction and feedback can invoke implicitly the generation of problem solving processes. It thus can be categorised as an informal learning environment for mathematics in the terms of this chapter since learning takes place through an inductive process.

A computer based environment does however have distinctive features which are relevant in terms of its potential for problem solving and project work. First, the computer can manipulate data.[3] This means that attention can be focused on the representation of the problem situation, the strategies of problem solution and the interpretation of the results rather than on formal manipulative techniques. Second, some software offers multiple representations of data and can provide different perspectives on the same problem. The opportunities for learning that these facilities open up need investigation. So the computer can be considered as a powerful tool within an informal learning environment. But, to simply think of the computer as a tool ignores its role as an expressive medium and can lead to a technocentric attitude to computers. According to Papert (1985) with reference to Logo:

"Technocentrism refers to the tendency to give ... centrality to a technical object – for example computers or Logo. This tendency shows up in questions like "what is THE effect of THE computer on cognitive development?" or "does Logo work? ... such turns of phrase often betray a tendency to think of "computers" and of "Logo" as agents that act directly on thinking and learning; they betray a tendency to reduce what are really the most important components of educational situations – people and cultures – to a secondary, facilitating role. The context for human development is always a culture, never an isolated technology. In the presence of computers, cultures might change and with them people's ways of learning and thinking. But if you want to understand the change, you have to center your attention on the culture – not on the computer." (his emphasis, Papert, 1987b, p. 25)

Thus, technocentrism implies a failure to notice that the computer is *not* just an efficient means of performing tasks and improving the users' performance, but also that it has the potential to transform the task itself and the relationship of the user to the underlying knowledge. Additionally technocentrism diverts attention away from the goal of the activity, here the mathematics, and away from the meaning of the activity for pupils.

To avoid technocentrism we must consider the nature of the software environments we wish to construct in terms of the means they provide for pupils to express themselves mathematically. If the software involves programming, interpreted in the widest sense, there is the possibility of user control both through the objectification of the tool, if this is useful for the development of mathematical knowledge, and through the construction of new tools. In the case of Logo, for example, procedures can be edited, new procedures written, named and used as building blocks within more complex programs. Programming software can thus form part of a rich (from the pupils' point of view) set of activities designed with the explicit intention to plant seeds for mathematical learning. Programming software is therefore at the heart of informal computer-based learning environments, abbreviated to I.CB.M. and called here mathematical microworlds.

COMPUTER-BASED MATHEMATICAL MICROWORLDS

Microworlds have been cited as examples of exploratory tools allowing pupils to investigate a representation of a given domain. They are seen as environments in which pupils can explore the consequences of the underlying logic, rules and relationships of the domain. Papert defined a microworld as: "...a slice of reality. It's strictly limited ... but it is rich ... the microworld is created and designed as a safe place for exploring ... and yet while being safe, it is also designed to be discovery – rich in the sense that little nuggets of knowledge have been scattered around in it for you to find" (Papert, 1987a, p. 79/80). He also suggested that within such a context children will learn to carry their personal explorations into the formal domain of scientific theory. Their creativity in this domain would not be stifled since the environment is "less dominated than the school's by the criteria of true and false" (Papert, 1980, p. 133).

We have defined the technical component of a Logo microworld for mathematics in more detail as:

"The most obvious feature of a Logo microworld is the software, the program or programs, which focus the learner's attention on specific ideas or processes. This is a set of tools which provides the learner with a representational system tailor-made for understanding a particular mathematical structure, in his or her own way. This technical component is composed, first and foremost, of the programming language. It is the programming language which provides the framework for the conceptual field modelled in the microworld.

By using computational tools with a focus merely on their output, pupils can be led to investigate the effects of changing initial values or testing the limits of applicability. However, the relationship to the programming language provides more: because of the way Logo is designed, it is reasonably straightforward to investigate the concept itself, and to dig down into the conceptual stack of ideas and relationships which form it. Pupils can look into the construction of the program, discriminate the effects of varying its constituent parts, and explore their relationship to the program as a whole". (Hoyles and Noss, 1987a, p. 587).

Thus we have attempted to provide pupils with opportunities to play within computer-based environments which have features which reflect the structure and composition of some embedded mathematics. Because of the extensibility available in programming environments, pupils can also create new tools for themselves.

For investigation within a microworld to be meaningful the activities and in particular the formalisations required, must "connect" with initial pupil conceptions and ways of working. Pupils do not come to school mathematics as 'empty vessels'. We know that pupils have well-established conceptions which frequently differ from those of the mathematician, and moreover persist despite conflicting evidence or teacher intervention. Pupils (and adults) tend to adopt an informal or intuitive perspective. diSessa (1982) in relation to physics describes the fundamental building blocks of such intuitions as 'phenomenological primitives'. He suggests that these 'can be understood as simple abstractions from common experiences that are taken as relatively primitive in the sense that they generally need no explanation; they simply happen ... the phenomenon is encoded simply as an expected event' (diSessa, 1985b, p. 4). In mathematics, the work of Hart (1981) and Booth (1984a) have illustrated the prevalence of 'incorrect' child methods. These may be personal and idiosyncratic, but, I would argue are, more often than not, correct within a limited domain of validity derived from use within earlier mathematical experiences, outside or within schools, and subsequently over-generalised. These child methods are not synthesised into any general theoretical framework. They appear as fragments called up by specific experiences, as 'knowledge in pieces' in diSessa (1988)'s terminology. The way to "connect" with pupil conceptions is therefore a piecemeal, complex and delicate task as discussed below.

A first level of complexity is that in designing mathematical microworlds it is important *not* to take pupil surface performance, particularly if derived from non-computer-based settings, as the starting point for two reasons. First, we need to build upon the pupil meanings of the mathematical knowledge and this may not match with their response behaviours. We have not found it helpful simply to identify pupil misconceptions, namely how and when they perform poorly or incorrectly, or how they use wrong methods. This prejudges intuitions from a standpoint of mathematical structures not yet assimilated in all their complexity. Rather, we try to design situations where pupils' potential can be released and realised. A similar plea was made by Booth (1984b) who suggested that, since after a short teaching experiment, children did actually improve their performance, the cognitive structures must have been available to the children. The only reason why, she argued, they had not previously applied these structures was the inappropriateness of the framework of reference in which they were working.

The following extract from a recent inverview undertaken as part of the pilot work in the Microworlds Project[4] serves as an example of this issue. In the Microworlds Project we have designed a computer-based microworld for Ratio and Proportion (Hoyles *et al.*, 1989). In a written pre-test, Stephen, a 12 year old boy, was required to construct responses within two contexts, one about mixing paint and the second about enlarging photographs of rugs. He *consistently* used an adding strategy (Hart, 1981). For example, Stephen gave the answer 16 to the following paint question:

"Jo mixes three tins of white paint with seven tins of black paint. Pat has twelve tins of white paint. Here is a table showing these amounts:

	Number of tins of white paint	Number of tins of black paint
Jo	3	7
Pat	12	?

How many tins of black paint must Pat use to get the same shade of grey?

On interview Stephen explained that he wrote 16 because "3 plus 9 gives 12 so you add 9 to the 7". This strategy was used *even when the answer was quite obviously ridiculous* as illustrated in his response to the following rug question:

"You collect photographs of rugs. You have just received a new set of photographs to add to your collection. You need to enlarge or shrink them to fit into the spaces in your catalogue. One of the 'new' lengths is unknown. Find the missing lengths (marked '?') on each diagram".

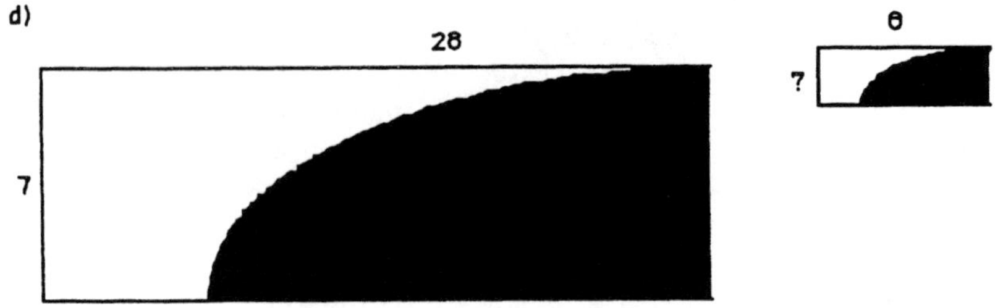

Missing length is ...

Stephen wrote the missing length as -13 arguing on interview "that it was 21 less than 8". It would have been all too easy to assume from his answers that Stephen's intuitions about the underlying structures of the problems were additive, but this was *not the case* as was evident from more careful probing. The reason why Stephen had used an additive strategy throughout the written test was that he recognised the *similarity* of the underlying structures of all the questions and *knew* therefore that there *had* to be a consistent procedure for answering all the questions. He chose addition merely because it was the only method he had readily available.[5] In searching for a consistent strategy across problems with similar structures, Stephen was in fact acting in a very mathematical way. What we have therefore to try to do is build on this mathematical intuition by providing an appropriate medium with which it can be expressed.

A second level of complexity in microworld design is that the computer environment may call up very different strategies from, for example, a paper and pencil environment. It is therefore important not to anticipate and prejudge pupil responses from a basis of past experiences in another medium. Settings can amplify or constrain activities (see for example Hoyles and Noss, 1989). Thus, as part of the microworld design process, there is a need for identification of pupil conceptions of the mathematical knowledge embedded in the microworld through analysis of their interactions within this microworld.

Another component of a mathematical microworld which needs to be built into its design, is the pedagogic component. If we assume that I.CB.M are simply examples of I.S.M then we have already identified the role of the teacher (see Table 1); namely to bring a mathematical perspective to the activity. However, if we do not make this assumption, we must attempt to define a distinctive role for the teacher within I.CB.M following the identification of a teaching role for the software; *a teaching role which is qualitatively different form explicit teaching and which arises from the interactive process and nature of the software tools used.* We will now consider these two issues in the following sections.

THE COMPUTER AS AN INTERACTIVE PARTNER IN THE DEVELOPMENT OF MATHEMATICAL KNOWLEDGE

In trying to understand pupil/computer interaction it is relevant to draw on work related to learning through social interaction for three reasons: first because of the social nature of mathematical knowledge (see the chapter by Balacheff in this volume), second because it has been argued that computer-based activities "invite" collaboration (Sheingold, 1987, p. 204) (although the circumstances in which this might happen need investigation[6]) and third, the reason most pertinent for the purposes of this chapter, in order to build up a basis for viewing the computer as one of the partners in the discourse.

In considering the role of peer group interaction in learning the importance of the complimentarity of points of view has been noted and how different points of view can provoke an elaboration of perspective together with reflection upon it and its possible modification (Mugny *et al.*, 1981; Light, 1983; Hoyles, 1985). Mugny *et al.* for example suggest that both social and cognitive conflict are necessary but not sufficient conditions for the efficacy of a social interaction in terms of the conceptual development of an individual. It is suggested that the following three conditions are likely to lead to "efficient" social conflict. First *participants are aware of the incompatibility of their responses; the conflicting response is comprehended and taken seriously* (see Balacheff chapter in this volume). This means that participants recognise that different responses *cannot* coexist and if responses are incompatible this incompatibility must be resolved. Second, *there are social criteria such as the presence of reciprocity between participants and the absence of dominance.* In a mathematical learning environment there is a third point additional to the general points above. *This concerns the 'correctness' of the pupil responses from a mathematical point of view.* This brings into the picture the role of materials and setting in providing feedback to pupils so that they can assess the correctness of responses against some outside yardstick.

So what is the role of the computer in relation to these three points? When the computer is one of the interactive partners, the first point, recognition of conflict, still remains an issue. How can pupils be provoked to reflect upon computer feedback and recognise that it might contradict their conjectures? Pupils have to learn to use the interactive process as a vehicle for theory building rather than merely the production of actions. The strategy of "try it and see", that is action based on immediate feedback, is often most efficient for production and certainly poses the least cognitive burden. We need to design situations in which pupils perceive the need to adopt a more analytic stance, making conjectures from a basis of mathematical structures. Pupils also have to be helped to see that the exactness of solution lies in the program and not the screen output (Kieran *et al.*, 1987).

The second point, the issue of reciprocity and group dominance, is relevant here if we interpret it broadly in terms of the important influence of emotional and affective issues. The way the computer is conceptualised is relevant and is beautifully described by Turkle (1987), 'The machine seduces you into psychologising and anthropomorphising it" (p. 54). We also must remember that the way pupils react to computers is affected by their experience of feedback devices and even by their keyboard skills. There are also people, often women, who are afraid of machines, afraid to break them and afraid when the machines respond to them, particularly when the responses appear out of control. We must remember that such people are likely to be dominated in any computer use.

As far as the third point is concerned, feedback needs to give information concerning the mathematical status of the pupils' actions in a way that does not inhibit further exploration, is understandable and possibly opens a door for an alternative pathway. A crucial notion here is the idea of a constructive bug in which "the learner gets enough feedback to recognise a bug, to determine its underlying causes and to develop procedures to correct the bug. This notion contrasts sharply with the notion of a non-constructive bug where a student may recognise that he is wrong but does not have the necessary information to understand why" (Papert, 1978, p. 148). The provision of the 'right' sort of feedback within the 'right' sort of activity is of course no easy task. An example of a 'quasi' constructive bug taken from the Logo Maths Project (Hoyles and Sutherland, 1989) is given below. The pupils developed a correct global intuition but did not have the local means of operation available to be successful since the computer tools were not adequate for this task.

Sally and Janet working with Logo had completed the ARROW task as given in Figure 1. They had built a procedure HILL with an input JACK where JACK represents the length AB. At the start of the episode the pair are trying to incorporate the length BE into their procedure. "Janet immediately suggested that they add another input for the new length BE.

Janet *"Yeh, so what we have to do is we have to make say JILL now"*

She then explained how she wanted to substitute JILL for all the 'bits' of the central stem of the ARROW.

Janet *"... so it's this one and this one and we have to change – so it's those two and those two we've got to change to JILL!"*

The pair agreed to this 'adding on' strategy for inputs[7] – that is every part of a figure which is perceived to vary is given a new input without explicit recognition of any internal relationship between the inputs. They typed:

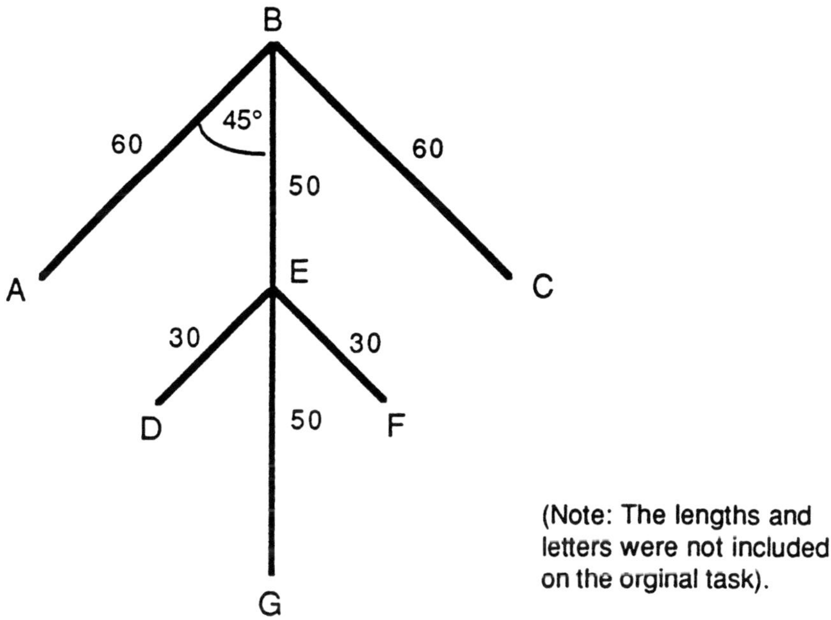

MAKE A PROCEDURE TO DRAW THIS SHAPE
AS BIG OR SMALL AS YOU WISH

Figure 1. The "Arrow" Task.

HILL :JILL :JACK

This formulation made Sally stop to reflect. She realised that *any* two numbers as inputs would not work!

Sally *"You can put any number in but it won't be the same..."*

At first Sally suggested subtraction as the relationship between JACK and JILL and Janet agreed – but at the point of writing this in Logo, that is as Janet proposed, FD SUB :JACK 10, Sally stopped again to reflect.

Sally *"Wait, if it gets bigger would it still be 10?"*

This is a critical observation. Sally seemed to be 'seeing the general in the specific' and recognising that the operation of subtraction would not work. Janet did not understand her point:

Janet *"... it would be right because you've got your hundred OK and you're subtracting 10 and you've got 90 ... you'll always be subtracting 10 'cos it's 10 less isn't it"*

This comment provoked Sally to try to explain her reservations and develop her argument more precisely in the face of continued opposition:

Sally *"... but if it's double as big then they would be double as small"*

Janet *"What?"*

Sally *"Right if this was 12 and it was subtract 10 then we'd only need 11 ... it wouldn't be the same"*

Janet *"It would"*

Sally *"It wouldn't 'cos ... say we subtract 1 from 6 would be 5 but then this would be 10 next time"*

Janet *"I dunno, I dunno what you're talking about..."*

Finally Sally was able to fully articulate her argument:

Sally *"Look this is 5 and if we made it twice as big that's 10 ... and this is 6 and if this was twice as big it would be 12 ... yeh" ... and if you said SUB : JACK 10 that would give you 11 and that's too long for that"*

At this stage Sally did not know what operation to propose but she was convinced that an additive strategy would *not* work. The fact that the inputs were declared on the turtle line of the Logo program and Sally knew that they could represent a whole range of numbers seemed to be important in this context. Janet however took over at this point and used subtraction since no other alternative was forthcoming.

At the end of the session the issue came up again as a result of the visual feedback which indicated that subtraction had *not* been the correct operation. Sally now knew that the operation required was multiplication but could not work out the number to use to multiply.

Sally *"There must be a number that you can do something with ... there must be some way to get the length of that the same as that ... I don't know that's what we've got to find out"*

On this particular occasion it proved too hard for the pair to work out the exact relationship required *but* the activity showed the *need* for a multiplicative relationship and the rejection of an additive strategy". (Hoyles and Sutherland, 1989, p. 190).

This computer-based activity together with the pair's discussion promoted the recognition of a bug in thinking about the underlying mathematics but fell short in the provision of the next step; that is assistance in the operationalisation of a method given a conviction as to its nature. How the computer environment can do this needs more investigation since the dialectical relationship of pupil negotiation and computer feedback and even how this varies according to task and computer environment is at present little understood.[8] However, as a starting point we will consider in the next section the notion of 'scaffolding' and how its domain of reference can be extended so as to form an explanatory variable in thinking about the ways mathematical knowledge can be developed in an informal computer environment.

PUPILS USING THE COMPUTER AS SCAFFOLDING

The notion of scaffolding is important in any consideration of informal learning environments. Scaffolding is a metaphor first coined by (Wood *et al.*, 1976) to describe support provided by a teacher or parent to a child which is "just enough" for their progression in the activity. It attempts to set out an ideal role for the teacher. Greenfeld (1984) points out that "The scaffold, as it is known in building construction, has five characteristics: it provides support, it functions as a tool; it extends the range of the worker; it allows the worker to accomplish a task not otherwise possible; and it is used selectively to aid the worker where need be ..." (p. 118). She goes on to suggest that "these characteristics also define the interactional scaffold provided by the teacher in an informal learning situation" (p. 118). A task is accomplished which, without the scaffolding, would not have been possible. In addition, Greenfeld argues there are more generalised effects beyond merely task completion and maintains "this process of collaborative work between teacher and learner often advances the learner's skills ..." (p. 118). With reference to Vygotsky's notion of the zone of proximal development (Vygotsky, 1978), he suggests that "adult guidance, when it functions like a scaffold in the zone of proximal development, not only leads the child to solve problems collaboratively that could not be solved alone, but also moves the embryonic skill towards its full blown manifestation"(p. 119). A phase in learning subsequent to scaffolding, and known as fading, is also important as Greenfeld later points out: "The scaffold not only helps successfully to accomplish the current task but also provides information which, as it becomes internalised, gradually eliminates the learner's need for the scaffold itself" (p. 124).

Now, let us look at this issue from a slightly different perspective of particular significance here. Bruner (1985) suggests that "the tutor in effect performs the critical function by 'scaffolding' the learning task to make it possible for the child, in Vygotsky's word, to *internalise external knowledge*" (my emphasis, p. 25). Thus, for our purposes we might ask the following questions: first, if 'external' mathematical knowledge is embedded in a microworld how can scaffolding assist its internalisation? Second, and following from the first, is it possible for the computer environment to provide scaffolding for the pupils *in ways not modelled on the idea of a tutor* (as in ICAI)[9] but in ways which are in the control of the learner and which arise from the *distinctive* features of the computer environment?

Scaffolding encourages children to perform actions which though recognisable and understandable to them, could not be performed without help. Thus scaffolding must be both pupil as well as task related. In talking about scaffolding it is important to emphasise that we are not interested in importing apprenticeship models of working into schools but, rather as Lave (1988b) suggests "because scenarios of apprenticeship learning are useful to 'think with'" (p. 2). In our case studies of pupils working in a Logo environment we have noticed scaffolding working in a distinctive way, erected and removed by the pupils; thus the pupils *themselves* find ways to exploit the computer facilities in a scaffolding role.[10] In turtle geometry, for example, we found that discrimination of significant features of an activity from a mathematical perspective is frequently fragmented. It is possible to discriminate the features of the graphical output without regard to numerical inputs or to the available symbolic representation, even when this symbolic representation, the program, is constructed by the pupil. It is also possible to discriminate within the symbolic representation and notice its structure and pattern but without regard to visual outcome. *However discrimination within one mode of representation can be used as scaffolding for discrimination in another mode.*

This is illustrated in the episode below taken from some case study work with 11 year old pupils (Hoyles and Noss, 1987b).

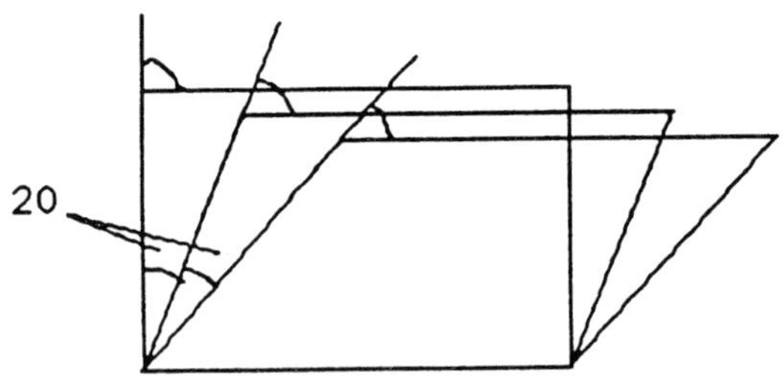

Figure 2. The Parallelogram Task.

Noel first wrote a procedure HOLO with two inputs defined as follows:

```
TO HOLO :R :Y
REPEAT 2[FD 200 RT :R FD 400 RT :Y]
END
```

When the sum of the two inputs to HOLO is 180, HOLO draws a parallelogram. Noel did not discriminate this necessary relationship explicitly yet he did use a correct pattern of inputs in building a procedure TII to draw the figure shown in Figure 2. TII was defined as follows:

```
TO TII
HOLO 90 90
RT 20
HOLO 60 120
RT 20
HOLO 30 150
END
```

TII did not work because Noel was not aware of the relationship between the input of 20 to RT (the interface between the HOLOs in TII), and the consecutive first inputs to HOLO that must be satisfied in order that all the parallelograms remain on the same base. On trying his procedure TII, he found that the bases of his parallelograms were moving. What is interesting here was that he eventually was able to work out what was wrong *by reflection on the visual effect of the different inputs.* Noel did not use an analytic approach to work out that the difference between consecutive first inputs to HOLO must be equal to the interface turn; the visual effects gave him the clues and provided the scaffolding. He commented to himself after working it out:

"It was not working last time 'cos I was not coordinating the angles properly. Everytime I took 20 off the angle I have to add it on to other".

Analysis similar to the above helps to identify those aspects of a computer environment which might help pupils through the 'hard parts of a task'. Taking another example, we know in mathematics that the transition from seeing a relationship between specific cases to writing it in explicit general form is not easy. Here the Logo environment can again serve as scaffolding. It allows pupils first to 'add on' variables as inputs to procedures simply in order to "flag" what is varying and sketch out a global view of what might be the significant features of any relationships. Pupils are thus able to capture the elements of a pattern which is "half" perceived but not yet fully comprehended. They can build up fragments and test each out phenomenologically, for example in Noel's use of the inputs to HOLO. By this means pupils can be relieved of the cognitive burden imposed by having to think about all the quantities and all the relationships within a figure at once.

The way partial ideas can be concretised in a computer environment, and made objects of reflection in the process of problem solution is illustrated in another experiment reported in Hoyles and Noss (1987b). Pupils were given a procedure SHAPE and asked to make up some tiling patterns using shape. SHAPE was defined as follows:

```
TO SHAPE :SIDE1 :SIDE 2
FD :SIDE1 RT 40
FD :SIDE2 RT 140
FD :SIDE1 RT 40
FD :SIDE2 RT 140
END
```

The following episode describes the work of Nicola.

"Nicola wanted to make a more general procedure to use for her tiling than SHAPE. She wrote the following procedure in the editor:

```
TO SHAPE1 :SIDE1 :SIDE 2 :MOVE
FD :SIDE1
RT :MOVE
FD :SIDE2
RT :MOVE
FD :SIDE1
RT :MOVE
FD :SIDE2
RT :MOVE
END
```

This procedure showed discrimination of the 'opposite lengths are equal' rule. A casual observer might assume that Nicola believed that all the turtle turns to form a parallelogram must be equal. However she immediately edited her procedure (*without running it*) to add an extra input MOVE1, which was then used as the input to the second and fourth RT commands. Her procedure was therefore as follows:

```
TO SHAPE1 :SIDE1 :SIDE2 :MOVE :MOVE1
FD :SIDE1
RT :MOVE
FD :SIDE2
RT :MOVE1
FD :SIDE1
RT :MOVE
FD :SIDE2
RT :MOVE1
END
```

Our interpretation of Nicola's work is that initially she did not want to focus on the turns but on the *overall program structure*; she simply wanted to put down a marker that they should vary. The computer allowed her to do this and then to return and 'correct the details'. The fact that her final program used the 'adding on' strategy also indicated that the embedded relationship between the angles was not yet extracted explicitly." (Hoyles and Noss, 1987, p. 153–154).

To give further illustration of the ways in which the computer can provide the scaffolding for pupils to find a way through a problematic step in a task when they have an *idea* for solution but are not yet in a position to *construct* all the steps, let us consider some more of Nicola's work in the same study. In trying to write a program for the figure given previously in Figure 1, she used the procedure for a rectangle, REC which she had constructed as follows:

```
"TO REC
 FD 100
 RT 90
 FD 200
 RT 90
 FD 100
 RT 90
 FD 200
 END
```

This procedure was not state-transparent and ended facing on a heading of 270 at A (as shown in Figure 3). After calling REC, Nicola typed in direct mode RT 135 FD 100 RT 45 FD 200 RT 135 FD 100 RT 45 FD 200, producing the screen output shown in Figure 3.

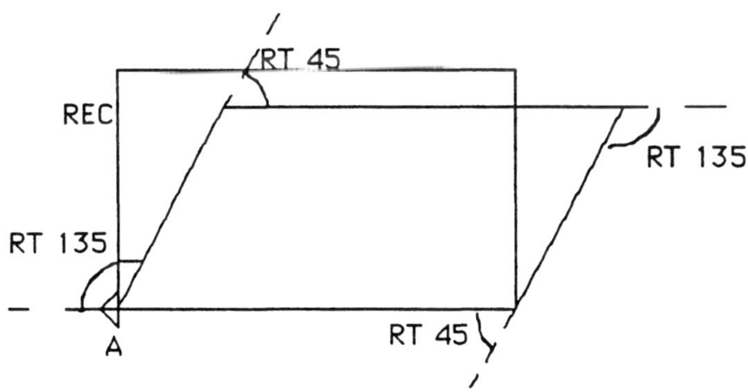

Figure 3. Nicola's First Attempt.

Nicola then put these commands following a call of REC in a super-procedure PAT as follows:

```
TO PAT
REC
RT 135
FD 100
RT 45
FD 200
RT 135
FD 100
RT 45
FD 200
END
```

She then (surprisingly) typed in the editor:

```
TO PAT1
PAT
RT :R
FD 100
RT :R
FD 200
RT :R
FD 100
END
```

Obviously the syntax for this procedure had quite a few bugs – but Nicola knew this! She again just seemed to want to get down *in an informal way* the shape of her ideas. Without running the procedure, she edited it as follows:

```
TO PAT1
PAT
RT 135 + 10
FD 100
RT 45 - 10
FD 200
RT 135 + 10
FD 100
RT 45 - 10
FD 200
END
```

At this point she tried out PAT1 and finally edited by 'tidying up' the inputs to the RT turns. We interpret her behaviour as follows: after seeing the pattern in the commands that worked, she recognised that the inputs to the turns must vary to produce the different parallelograms – so she put them all down as :R. Then, having found a pair of turns that worked (135 and 45), she hypothesised that she must add 10 to her first turn and subtract 10 from the second to construct the next parallelogram. Her final procedure for the given pattern continued with this 'adding and subtracting 10' strategy as follows:

```
TO PAT1
PAT
RT 145
FD 100
RT 35
FD 200
RT 145
FD 100
RT 35
FD 200
RT 155
FD 100
RT 25
FD 200
RT 135
FD 100
RT 25
FD 200
END   "(Hoyles and Noss, 1987b, pp. 155–157)
```

We interpreted this episode by suggesting "that the computer allowed Nicola to set up a global symbolic framework, then focus on the local components she was trying to work out, and adopt a step-by-step strategy with respect to these components. Thus she was able to work on a series of specific cases and use these as a way to move towards a general strategy." (Hoyles and Noss, 1987b, p. 157)

This is an example of a computer environment providing scaffolding used under pupil control. The pupil actions combined with the visual feedback were 'ahead' of the vocabulary available to describe the activity. This 'gap' was filled temporarily by the computer. Not only did Nicola scaffold the task for herself but also she organised her own fading, that is the removal of scaffolding support. *Thus the computer environment can allow critical features of activity to be marked, adjusted and then eventually built into a mathematical structure.* Although it is evident that the computer here did provide a teaching role in the development of mathematical knowledge, we need to think through carefully the interrelationship between this role and the role of the teacher. Would for example Nicola, without teacher intervention, have made the relationship explicit between the two inputs MOVE and MOVE1 in her program in the first episode? Or, more generally, would she have removed the scaffolding of marking each turn and working out a relationship between the *numbers* reqûired for the inputs and build this relationship in terms the *symbols* representing the numbers. We have seen some pupils dispose with the "adding on" strategy and simply define the minimum necessary inputs together with the appropriate explicit relationships in the program. They thus choose to cast aside the computer's scaffold of 'redundant' inputs. This does *not* however necessarily happen. The circumstances in which this type of progression does and does not take place need more precise analysis.

CONCLUSION

To try to draw together the arguments presented in this paper, I will return to my original analysis of informal learning environments but now introduce into this analysis informal computer-based learning environments for mathematics (I.CB.M). I.CB.M and I.S.M. have many facets in common and it is worthwhile making explicit these similarities, as follows:

- the mathematics is embedded and contextualised (and usually messy)
- there is a product which is concrete and/or visualisable, thus available for inspection
- the tools available are defined by the setting
- collaborative modes of learning are frequently adopted
- the problems/tasks (at least partly) are organised and/or controlled by the problem solvers
- the rationale for the tasks is to provide vehicles for generalisations and their formalisation.

So what are the differences between I.CB.M and I.S.M? The major difference is that within I.CB.M the pupils have experiences of 'playing' with formal systems, and are able to link their intuitions to the generalisation process. The teacher's role in highlighting the mathematical content can therefore build on the pupil activity rather than distort it. It is often assumed that the manipulation of concrete materials in schools necessarily involves an explicit sense of the manipulation of the underlying mathematical structures and relationships. This is evidently *not* the case. However the manipulation of objects which carry with them quantities or symbolisation provides a rather different scenario and can provide the link between using mathematics informally and its discrimination and generalisation. Thus in contrast to I.S.M., I.CB.M. can sensitise the user to mathematical relationships implicit in the activity, by means of interaction, feedback and scaffolding.

Within I.CB.M, we must be able to control and explore the software itself. The software cannot therefore merely consist of 'black boxes' whose internal workings and limits of applicability are not in the users' control. As diSessa (1985) puts it 'Expressive power and nuance are incompatible with invisibility and inaccessibility' (p. 2).[11] This argument implies that the software involves some sort of programming. Programming activity requires a focus on the *process* of construction of a product and *its representation in a generalised way in the program.* It is these two programming requirements which are *the features of I.CB.M. which mark them out as distinctive informal learning environments for the development of mathematical knowledge.* It is these features I suggest which have the potential *to bridge the gap between fragmentary mathematical actions within informal environments and their incorporation into a mathematical system.* I.CB.M, unlike I.S.M, can therefore *promote the transition from the primacy of the relationship of the mathematical tools to the environment to the primacy of the relationship between the mathematical tools themselves.*

It is envisaged that in school the teacher will remain the pivotal mediator of the technicalities of the software and the connections between the computer work and the paper and pencil work. The teacher will provide also the subtle metacognitive and affective feedbacks essential for learning. The role of the teacher in I.CB.M. is therefore crucial for both managing the complexities of the situation and promoting learning. Pupils tend not to feel the 'need' to push towards generalisation, a practice so fundamental for the mathematician. So the teacher has a central role in helping the pupils to see the "pay off" of leaving their naive empiricism and moving up a level in abstraction. This we argue will happen more naturally within I.CB.M. We thus see the computer medium as augmenting the teacher by providing an exploratory problem-solving environment which has an instructional as well as interactive role in the learning process.

To summarise, I have compared informal everyday mathematics with school mathematics and pointed to ways in which informal computer-based environments can capture the power and motivation of the former activities for development in a school setting. Computer tools can be used to produce particular products and can generate problem solving activity and genuine pupil decision making. Additionally, if the software involves programming of any kind, precise formalism is required. It is this formalism which can bridge the gap between pupils actions structured by the goals of their project and pupil actions structured by mathematics (provided that the formalisation arises from and matches the pupil actions (see Hoyles and Sutherland, 1989)). This happens in two ways: First the formalisation performs the function of highlighting the embedded mathematical structures and relationships. Second, the formalisation captures the generalisation which can then be used as a vehicle to assist in the transition from seeing specific cases *only* as such to seeing specific cases as generic examples. The software acts as a partner with the pupil, a partner who interacts and responds but who also can provide stepping stones from mere use of mathematical ideas to their discrimination and generalisation. Multiple representations in which discrimination in one representation can provide scaffolding for discrimination in another prior to a synthesis commensurate with more robust understanding is important here. Thus the potential of the computer to provide scaffolding towards the generalisation of specific cases and towards the synthesis of the mathematical fragments encountered in informal environments has formed the basis of the argument. This teaching role for a computer-based environment is seen as qualitatively different from explicit teaching and complementary to it. We need new ways of thinking about the development of mathematical knowledge through computer-based microworlds. This is a beginning perhaps but there is a long way to go!

ACKNOWLEDGEMENT

I would like to thank Richard Noss for his helpful comments on an earlier draft of this paper.

NOTES

[1] As Resnick (1987) suggested 'Both the structure of knowledge used and the social conditions of its use may be more fundamentally mismatched than we previously thought' (p. 15).
[2] For a discussion of the devolution process in classrooms, see Brousseau (1988).
[3] Algebraic, functional and graphical manipulations are implied here as well as the numerical.
[4] The Microworlds Project is a Research Project codirected by the author with Richard Noss and Rosamund Sutherland.
[5] The questions 'required' a range of scale factors from integer to fractional and he was unsure of the *method* of multiplying by a fractional quantity.
[6] This is the subject of a Research Project "Groupwork with Computers" under the joint directorship of the author and Professor M. Eraut of University of Sussex, U.K.
[7] Hoyles (1987) noted that children in the early stages of writing Logo procedures tend to append an input to the title line of a procedure for every entity that is variable. They thus circumvent the need to define any explicit relationship between the inputs within the program – although the relationship is understood frequently at an implicit level as shown by the values chosen for the inputs which adhere to this relationship.
[8] This is the subject of an ongoing research project funded by the Leverhulme Trust with Rosamund Sutherland and Lulu Healy. In this project we are investigating how pairs of pupils come to make generalisations in activities using Logo, Spreadsheets (EXCEL) and paper and pencil.
[9] ICAI Intelligent Computer Assisted Instruction.
[10] A familiar analogy would be the way people use word processing facilities to help them structure their writing.
[11] A related point is made with respect to invisible mathematics by Keitel (1989).

REFERENCES

Booth, L.: 1984a, *Algebra: Children's Strategies and Errors*, NFER-Nelson, Windsor.

Booth, L.: 1984b, 'Misconceptions leading to Error in Elementary Algebra', *Journal of Structural Learning* **8**, 125–138.

Brousseau, G.: 1988, 'Basic Theory and Methods in the Didactics of Mathematics', *Proceedings of the 2nd Conference on Systematic Cooperation between Theory and Practice in Mathematics Education*, Verstappen, P. (ed.), National Institute for Curriculum Development.

Bruner, J.: 1985, 'Vygotsky: An Historical and Conceptual Perspective', in Wertsch, J.V., (Ed.), *Culture, Communication and Cognition: Vygotskian Perspectives*, Cambridge University Press, NY, USA.

Carraher, T.N., Carraher, D.W., and Schliemann, A.D.: 1985, 'Mathematics in the Streets and in Schools', *British Journal of Developmental Psychology* **3**, 22–29.

Carraher, T.N., Schliemann, A.D., and Carraher, D.W.: 1988, 'Mathematical Concepts in Everyday Life', in Saxe, G. and Gearhart, R.M., (eds.), *Children's Mathematics. New Directions in Child Development*, San Francisco, Jossey Bass, pp. 71–88.

diSessa, A.: 1982, 'Unlearning Aristotelian Physics: A Study of Knowledge-based Learning', *Cognitive Science* **6**, 37–75.

diSessa, A.: 1985, *Notes on the Future of Programming: Breaking the Utility Barrier*, MIT Occasional Paper.

diSessa, A.: 1988, 'Knowledge in Pieces', in Forman, G. and Pufall, P. (eds.), *Constructivism in the Computer Age*, Lawrence Erlbaum Associates, Hilsdale New Jersey.

Gerdes, P.: 1988, 'A Widespread Decorative Motif and the Pythagorean Theorem', *For the Learning of Mathematics* **8**, 35–60.

Greenfield, P.M.: 1984, 'A Theory of the Teacher in the Learning Activities of Everyday Life', in Rogoff, B. and Lave, J. (eds.), *Everyday Cognition: Its Development in Social Context*, Harvard University Press, Cambridge, Massachusetts, pp. 117–138.

Harris, M.: 1987, 'An Example of Traditional Women's Work as a Mathematics Resource', *For the Learning of Mathematics* **7**, 26–29.

Hart, K. (ed.): 1981, *Children's Understanding of Mathematics*, John Murray, London.

Hoyles, C.: 1985, 'What is the Point of Group Discussion in Mathematics?', *Educational Studies in Mathematics* **16**, 205–214.

Hoyles, C.: 1986, 'Scaling a Mountain – A Study of the Use, Discrimination and Generalisation of some Mathematical Concepts in a Logo Environment', *European Journal of Psychology in Education* **1**, 111–126.

Hoyles, C.: 1987, 'Tools for Learning – Insights for the Mathematics Educator from a Logo Programming Environment', *For the Learning of Mathematics* **7**, 32–37.

Hoyles, C. and Noss, R.: 1987a, 'Synthesising Mathematical Conceptions and Their Formalisation Through the Construction of a Logo-based School Maths Curriculum', *International Journal of Mathematical Education in Science and Technology* **18**, 581–595.

Hoyles, C. and Noss, R.: 1987b, 'Children Working in a Structured Logo Environment: From Doing to Understanding', *Recherches en Didactique de Mathematique* **8**, 131–174.

Hoyles, C. and Noss, R.: 1989, 'The Computer as a Catalyst in Children's Proportion Strategies', *Journal of Mathematical Behaviour* **8**, 53–75.

Hoyles, C. and Sutherland, R.: 1989, *Logo Mathematics in the Classroom*, Routledge, London and New York.

Hoyles, C., Noss, R., and Sutherland, R.: 1989, 'Designing a Logo Based Microworld for Ratio and Proportion', *Journal of Computer Assisted Learning* **5**, 208–222.

Kieran, C., Hillel, J., and Gurtner, J.L.: 1987, 'Qualitative Strategies in Logo Centering Tasks', *Proceedings of Second Logo and Mathematics Education Conference*, University of London Institute of Education, pp. 126–138.

Keitel, C.: 1989, 'Mathematics, Education and Technology', *For the Learning of*

Mathematics **9**, 7–17.

Lave, J.: 1988a, *Cognition in Practice: Mind, Mathematics and Culture in Everyday Life*, Cambridge University Press, Cambridge.

Lave, J.: 1988b, *The Culture and the Practice of Understanding*, IRL Report No. IRL88–0007, May, Institute for Research on Learning, California.

Lawler, R.W.: 1985, *Computer Experience and Cognitive Development: A Child's Learning in a Computer Culture*, Ellis Horwood, Chichester, U.K.

Light, P.: 1983, 'Social Interaction and Cognitive Development: A Review of Post-Piagetian Research', in Meadows, S. (ed.), *Developing Thinking*, Methuen, London.

Mellin-Olsen, S.: 1987, *The Politics of Mathematics Education*, D. Reidel Publ. Co., Dordrecht, Holland.

Mugny, G., Perret-Clermont, A.N. and Doise, W.: 1981, 'Interpersonal Co-ordination and Sociological Differences in the Construction of the Intellect', in Stevenson, G. and Davis, G.B. (eds.), *Applied Social Psychology*, Vol. 1., Wiley, Chichester.

Papert, S.: 1978, *Computer-based Microworlds as Incubators for Powerful Ideas*, MIT, A.I. Laboratory.

Papert, S.: 1980, *Mindstorms: Children, Computers and Powerful Ideas*, Harvester Press, Brighton, U.K.

Papert, S.: 1987a, 'Microworlds: Transforming Education', in Lawler, R.W. and Yazdani, M., *Artificial Intelligence and Education*, Vol. 1, New Jersey: Ablex, pp. 79–94.

Papert, S.: 1987b, 'Computer Criticism *vs* Technocentric Thinking', *Educational Researcher* **16**, 22–30.

Resnick, L.B.: 1987, 'Learning in School and Out', *Educational Researcher* **16**, No. 9, pp. 13–20.

Sheingold, K.: 1987, 'The Microcomputer as a Symbolic Medium', in Pea, R.D. and Sheingold, K. (eds.), *Mirrors of Minds: Patterns of Experience in Educational Computing*, Ablex Publishing Corporation, New Jersey, USA, pp. 198–208.

Sutherland, R.: 1988, 'What Are the Links Between Variable in Logo and Variable in Algebra?', *Recherches en Didactiques de Mathematiques* **8**, 103–129.

Turkle, S.: 1987, 'Computational Reticence: Why Women Fear the Intimate Machine', in Kramerae, C. (ed.), *Technology and Women's Voices?*, Routledge & Kegan Paul, New York & London.

Vygotsky, L.S.: 1978, *Mind in Society: The Development of Higher Psychological Processes*, Harvard University Press, Cambridge, Massachusetts.

Walkerdine, V.: 1982, 'From Context to Text: A Psychosemiotic Approach to Abstract Thought', in Beveridge, M., *Children Thinking Through Language*, Edward Arnold, London, pp. 129–156.

Wood, D., Bruner, J. and Ross, G.: 1976, 'The Role of Tutoring in Problem Solving', *Journal of Child Psychology and Psychiatry* **17**, 89–100.

ENTR'ACTE 8

From the new world of the computer microworld we return to a more familiar world of the mathematics classroom – but to a different and newer exploration of its possibilities. Balacheff concentrates on the important topic of 'mathematical proof' but not the dry, 'finished-product' version as represented in the formal mathematical text. Balacheff's chapter leads on nicely from Hoyles' with the common thread being that of discussion, of negotiation and of social construction. Balacheff analyses the potential development of proof through social interaction in the mathematics classroom. What is it that is gained by the learners in such an interaction, but what also is lost? And how, crucially, do they behave in such a situation? Werner's chapter has already made the reader aware of the richness of children's ideas; what then will they bring to problems involving rational argument and conviction which are established in a social interaction situation? How can a teacher retain control of the social interaction so as to maximize its benefits while minimizing the effects of its weaknesses?

THE BENEFITS AND LIMITS OF SOCIAL INTERACTION: THE CASE OF MATHEMATICAL PROOF

INTRODUCTION

Teaching mathematical proof is a goal in most mathematical curricula. This target is sometimes explicitly stated, that is "mathematical proof" can appear as a chapter in the textbook, or it is sometimes expressed as a general aim of the teaching. In this latter case, the aim is the training in the construction and the formulation of deductive reasoning, "mathematical proof" not being named as such. Actually, that is the situation at the present time in France: the last French programmes[1] (1985) for the eighth grade state that students should be trained progressively to construct deductive reasoning, and this training must bc intensive in the ninth grade. The situation is the same in other countries, like in Québcc:

"To develop pupils' logical thinking by bringing them to structure their reasoning [...] It is therefore advisable to demand that the pupils' problem-solving is based on logical and structured reasoning."
(Objectif du programme[2] du secondaire V, Québec, 1984)

The situation is likely to be the same in US, following the indication of the recent Standards for School Mathematics:

"All students, especially the college intending, should learn that deductive reasoning is the method by which the validity of a mathematical assertion is finally established" (NCTM, 1989, p. 143).

This objective is very difficult to fulfil according to teachers (Gaud and Guichard, 1984), and as it is established by research (Bell, 1976, Senk, 1985; Usiskin, 1982). Usually two reasons are offered by mathematics teachers to explain these difficulties, one is the lack of students' awareness of the necessity to give any proof, the other is their lack of logical maturity. But, in my opinion, these explanations are rather inadequate insofar as outside the mathematics classroom there is some evidence that the same students behave quite a differently. To understand what happens we have to enter the students' world, and we have to agree to consider the coherence of their own rationales for asserting the truth of a mathematical statement or a result. In other words, we must make our own Copernican revolution. We will

Alan J. Bishop et al. (eds.), Mathematical Knowledge: Its Growth Through Teaching, 175–192.

discover that students have some awareness of the necessity to prove and some logic (Balacheff, 1988). Perhaps these are not identical to what we expect them to be. The related open question is "why?" I would like to suggest that the answer to this question cannot be found just by considering the student in isolation. I mean that we have to consider the context in which s/he behaves: students' cognitive behaviors are meaningful insofar as they are *their* answer to the situation as *they* perceive it.

Some research (Schoenfeld, 1988) has demonstrated that the teaching situation in which students construct the solution to a given problem and in which they formulate it, can lead them to radically transform their initial solution, and the means used to validate it, into something they find more suitable to the teacher expectations as they understand them. For example, a student I observed[3] while he was solving a problem together with one of his school mates, after having found a solution in quite a good manner, then engaged in an odd process just to formulate his solution mathematically. Unfortunately, the solution he then produced no longer looked like anything coherent (Balacheff, 1982). As Schoenfeld (1988) states: "For these students failing to use the proper form – whether or not the substance of what you have produced is correct – is being unmathematical". We must realize that in such a situation teachers have no way to think anything else than that such students behave at a very low level of logic. This phenomenon is now well known to many researchers, and there is a consensus to consider that "students' apparently bizarre [mathematical] behaviors frequently cannot be accounted for solely in terms of conceptual limitations. [There] is a need to move beyond the 'purely cognitive'." (Cobb, 1986, p. 2).

What most of the researchers discover, when moving beyond 'purely cognitive', is the crucial role played by social issues. As, for example, Lampert stated:

> In the mathematics classroom "doing mathematics means following the rules laid down by the teacher; knowing mathematics means remembering and applying the correct rules when the teacher asks a question; and mathematical truth is determined when the answer is ratified by the teacher." (Lampert, 1988, p. 437).

These characteristics of school mathematics are of a social nature. Recognizing this, Lampert then reports that "changing students' ideas about what it means to know and do mathematics was a matter of immersing them in a social situation that worked according to different rules than those that ordinarily pertain in classroom, and then respectfully challenging their assumptions about what knowing mathematics entails." (*ibid.*, p. 470). Such a project is at the core of many attempts to modify classroom practices in order to obtain a modification of students' view about what mathematics consist of (e.g. Arsac et Mante, 1983; Capponi, 1986). Schoenfeld (1990) explored the issue of mathematics as "sense-making" and showed that this kind of social consideration is not a recent development, as for example we can find classroom discussions promoted by Fawcett (1938). What is new now, seems to be the consensus of the mathematics educators' community about the fact that issues related to the social dimension must be considered much more carefully. Evidence of the fruitfulness of using social settings to facilitate mathematics learning consist mainly of stories about the success of such attempts. But, as we know such settings do not work with every teacher or in every classroom. Research is needed to strengthen the foundation of such approaches. Part of my own research has been to explore the necessary conditions for the success of such settings and thus to explore both benefits and limits of the use of social interactions in the mathematics classroom. It is what I will present now, focusing on the case of teaching and learning mathematical proof.

THE MEANING OF MATHEMATICAL PROOF: AN OUTCOME OF SCHOOL PRACTICE

What a mathematical proof consists of seems clear to all mathematics teachers and mathematics educators. That is: "A careful sequence of steps with each step following logically from an assumed or previously proved statement and from previous steps." (NCTM, 1989, p. 144). This description is almost the same all over the world, and it is very close to what a logician would formulate, perhaps more formally. Comments made on mathematical proof as a content to be taught emphasize two points: first they stress that it has nothing to do with empirical or experimental verification, second they call attention to the move from concrete to abstract. Here are examples of such comments:

"It is a completely new way of thinking for high school students. Their previous experience both in and out of school has taught them to accept informal and empirical arguments as sufficient. Students should come to understand that although such arguments are useful, they do not constitute a proof." (NCTM, 1989, p. 145).

We can say that the definition of mathematical proof, as an outcome of these official texts is mathematically acceptable, but there is a long way between this definition and the students' concept-image of mathematical proof as a result of teaching interactions. More or less, teaching mathematical proof is understood as teaching how to formulate deductive reasoning: "For teachers, a mathematical-proof[4] is very clearly the formal deductive exposition of logical reasoning" (Braconne, 1987, p. 187). What Schoenfeld (1988), in the case of high school geometry classroom in USA, points to is "a strict protocol for writing proofs" that students are expected to follow closely.

The construction of this reasoning, and its possible relationships with other kinds of reasoning, is hidden by that over-emphasis on its "clear" formulation. That conception is so strong that some teachers can come to evaluate a mathematical proof by just considering the surface level of the discourse. For example, in her requirement for teachers' comments on a sample of students formulations, Braconne reports[5] that:

"The teachers reacted to the useless length of Bertrand's answer, to the disorderliness in Karine's solution, to the fact that Elodie's answer did not follow deductive reasoning, etc. Nevertheless, seven teachers did not notice that in Bertrand's answer the equivalent of the required theorem was mentioned in the first paragraph, and eight did not point out that Laurent's answer contained the same error [...] So for the student, and for us, the marks given did not reflect whether the teacher had noticed the error or not." (Braconne, 1987, p. 99).

A report on proof frames of elementary preservice teachers shows a similar behavior:

"Many students who correctly accept a general-proof verification did not reject a false proof verification; they were influenced by the appearance of the argument – the ritualistic aspects of the proof – rather than the correctness of the argument. [...] Such students appear to rely on a syntactic-level deductive frame in which a verification of a statement is evaluated according to ritualistic, surface features." (Martin and Harel, 1989).

Thus, mathematical proof appears ultimately as a kind of rhetoric specific to the mathematical classroom. It is not surprising then that it appears as such to the eyes of students. The nature of a mathematical proof as a tool to establish a mathematical statement is to some extent hidden by the emphasis on the linguistic dimension. What does not appear in the school context is that a mathematical proof is a tool for mathematicians for both establishing the validity of some statement, as well as a tool for communication with other mathematicians.

Beyond a certain complexity, conceptual or technical, how can we come to certainty about the validity of the discourse which intends to guarantee the truth of a mathematical result? The precise "symbolic technology" (Bishop, 1988) used to express a mathematical proof is one of the means to shake that validity, but as Manin[6] recalls, ultimately "a proof becomes a proof after the social act of 'accepting it as a proof'." This social nature of mathematical proofs is part of their practical value, leading to acknowledging them as reliable and efficient tools for the mathematician. Actually, we should remember that what constitutes the present consensus about rigor has not been created *ex nihilo*, it is the product of an historical and a social process within the community of mathematicians.[7]

There is another reason for considering strongly the social dimension of mathematics teaching and learning. We have chosen to work with the constructivist hypothesis which asserts that the subject explores actively his or her environment, that he or she participates actively in the creation of space, time and causality (Inhelder and Caprona, 1985, p. 8). So we recognize learning as a personal process, and we thus consider that its outcome is likely to be firstly a private knowledge: the students' conceptions. But that conflicts with two constraints specific to the teaching, which has to guarantee the socialization of students' conceptions for at least the two following reasons:

- *Mathematics is social knowledge.* It has a social status in society, or in smaller social groups under whose control it is used. That means that students should make their own knowledge that which exists outside the classroom; they are not free from the meaning they construct. For example, the community of mathematicians or that of engineers can be taken as a social reference.
- *The mathematics class exists as a community.* The teacher has to obtain a certain homogeneity in the meaning of the knowledge constructed by students, and she or he has to ensure its coherence. Otherwise, the functioning of the class will be hardly possible.

Because of the constructivist hypothesis, we consider the use of authority is not desirable. Thus the homogenization can only be the result of a negotiation within the classroom, or of other specific social interactions such as the one Brousseau (1986) has described in the frame of his *théorie des situations didactiques*.

SOCIAL INTERACTION AND SITUATIONS FOR VALIDATION

What is now clear is that as long as students rely on the teacher to decide on the validity of a mathematical outcome of their activity, the word 'proof' will not make sense for them as we expect it to do. In such a context they are likely to behave mainly to please their teacher, just as one of the British students interviewed by Galbraith (1979) told his interviewer: "To prove something in maths means that you have worked it out and it proves how good you are at working questions out and understanding them."

But it is not sufficient to propose a problem to the mathematics class, and to tell the students that they have the responsibility for solving it. There is no reason for them, *a priori*, to consider that that problem is *their* problem and to feel committed in solving it; they can still think that they have to do so in order to please the teacher and thus their behavior will not be very significant.

Before going ahead, let us consider a short story told by Sir Karl Popper, it will throw a relevant light on what we want to suggest:

"If somebody asked me, 'are you sure that the piece in your hand is a tenpenny piece?' I should *perhaps glance at it again* and say 'yes'. But should a lot depend on the truth of my judgement, I think I should take the trouble to go into the next bank and ask the teller to look closely at the piece; and if the life of a man depended on it, I should even try to get to the Chief Cashier of the Bank of England and ask him to certify the genuineness of the piece." (Popper, 1979, p. 78).

Then Popper adds that "the 'certainty of a belief is not so much a matter of its intensity, but of the *situation*: Of our expectation of its possible consequences." (*ibid.*)

On the same line, I would like to suggest that if students do not engage in any proving processes, it is not so much because they are not able to do so, but rather that they do not see any reason. Even if they engage in such a process, its nature depends heavily on the way students understand the situation. Following a principle of *economy of logic* they are likely to bring into play no more logic than what is necessary for practical needs (Bourdieu, 1980, p. 145).

Then the true meaning of the outcomes of students proving processes, and even the fact that they enter or not into such a process, is to be traced in the characteristics of the situation in which they are involved.

In situations in which they have to decide a common[8] solution to a given problem, students have to construct a common language and to agree on a common system to decide on the validity of the solution they propose. The essential role of the social dimension, mainly in situations for communication, is in provoking a move from "doing" to "telling how to do". Their importance in the construction of meaning has been demonstrated by Brousseau in his *théorie des situations didactiques* (Brousseau, 1986). Here we would like to recall what this author wrote about the situations for validation:

The situations for validation "will bring together two players who confront each other regarding a subject of study composed on the one side of messages and descriptions produced by the pupils and on the other side of the a-didactic milieu used as referent for these messages. The two players are alternately a 'proposer' and an 'opposer'; they exchange assertions, proofs and demonstrations concerning this pair 'milieu/message'. This pair is a new apparatus, the 'milieu' of the situation for validation. It can appear as a problem accompanied by the attempt at solving it, like a situation and its model, or like a reality and its description ...
While informer and informed have disymmetric relations with the game (one knows something that the other does not know), the proposer and the opposer must be in symmetrical positions, both as regards the information and means of action about the game and the messages which are at their disposal, and as regards their reciprocal relations, the means of sanctioning each other and the objectives vis-à-vis the pair milieu/message." (Brousseau, 1986, p. 158).

We should realize that in such situations, behaviors that are more social than mathematical, would probably appear. For example, because of self-esteem, some students might refuse to recognize that they are wrong, or others might refuse to accept that their opponents are right.

Thus, to provoke students' proving behaviors one of the possible solutions is to design situations in such a way that students come to realize that there exists a risk attached to uncertainty, and thus that there is an interest in finding a good solution. In order to obtain a significant scientific debate among students, we should provide them with a situation promoting contradiction, but also promoting acceptance. Otherwise systematic rejection could become an efficient defensive strategy. In other words the situation should allow the recognition of a risk linked to the rejection of a true assertion, or to the acceptance of a false one.

Following these principles we have designed teaching situations as experimental settings in order to study students' behaviors in such contexts, and the nature of these behaviors in relation to the characteristics of these situations. *A priori*, we thought that genuine mathematical proving processes will be observed, but a deeper analysis of our experiments has shown that things are more complex than what is usually acknowledged by innovators who base their work on the use of social interaction in the classroom.

In the following section we will report in some detail on one of these experiments.

EVIDENCE FROM A CASE STUDY[9]

A first principle we wanted to satisfy in designing the experiment was to obtain the devolution[10] of the responsibility for the validity of the problem's solution from the teacher to the students. For that purpose we chose a context of communication: We told the students that they will have to write a message to other students, of the same grade, in order to allow them to solve a given problem. In such a situation the criteria for success are left to be decided to the students, according to their own means for the evaluation of the efficiency and of the reliability of the message they have produced. We thought that this setting would be sufficient to ensure that students will consider that they have the responsibility to decide on the truth of their solution, and without any concern about the teacher expectations.

In such a situation there are usually some tensions because of the different individual motivations and commitment. For this reason, we think that it is not desirable to ask the students to work individually. On the other hand it is not desirable to ask for a collective production from the whole class insofar as some students might feel that they are not concerned, leaving the task to the others. So, we decided to constitute small teams of three to four students working together, telling them that the final solution will be one of the ones proposed by the teams, or a modification of it. To promote collective work, each team had to propose only one solution. During the debate for the choice of the class solution the team was asked to express its position through the voice of a chosen representative. That constraint obliges students to explicitly discuss *a priori* the goodness and appropriateness of what they want to be said. We think that the quality of the debate will rest on the motivation of each team, its willingness to have its message chosen, but also its commitment to the success of the class as a whole.

The mathematical problem we chose was the following:

> *Write for other students, a message allowing them to come to know the perimeter of any triangle a piece of which is missing. To do it, your colleagues will have at their disposal only the paper on which is drawn a triangle and the same instruments as you (rulers, etc.).*

Together with this text a triangle like the following (Figure 1) was given to the students. All the teams in the classroom had the same materials.

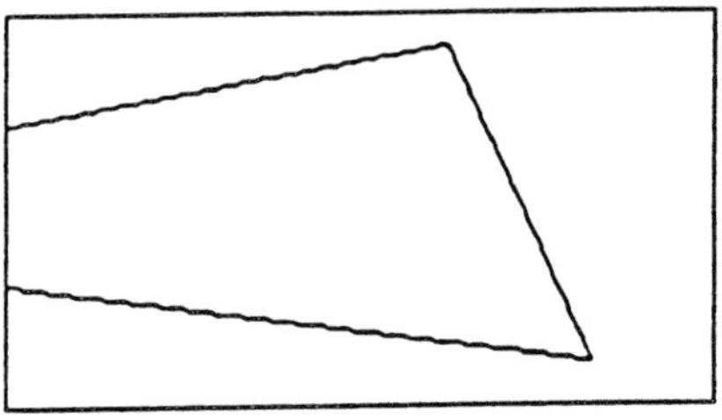

Figure 1.

A study we made before this experiment (Balacheff, 1988, pp. 321–360), allowed us to think that all the students will be able to enter the problem-solving process, with quite different solutions. This diversity was expected to be the source of interesting debates. We know that some students, and thus some teams, will miss the fact that the solution must work for a general case and not only for the triangle given as an example. But we were sure that this will be pointed out during the debate, and then that it will be taken into consideration, even with more strength than if the teacher had warned about it *a priori*.

The role of the teacher was to present the situation, and not to intervene in any case until all the teams had proposed a solution. Then her position would be to regulate the debate and to give the floor to the teams' representatives. The end of the sequence will come from a general agreement on the fact that one of the solutions, or a new one obtained as a result of the interactions, is accepted. The debate was organised in the following way: the messages were written on a large sheet of paper and were displayed on the wall of the classroom. Each team had to analyze the messages and their representatives had to tell the class their criticisms and suggestions. These criticisms had to be accepted by the team whose message was discussed. In case of an agreement of the class on a false solution, the teacher was allowed to propose a new triangle to the teams, invoking that such a triangle might be considered by the recipients (such material had been prepared taking into account what we learned from the first study). On the other hand if more than one message is acceptable with no clear decision from the class, then the teacher is supposed to organise a vote to make the choice, asking the students to tell the reasons for it.

I will not here report in details on the analysis of this experiment, a complete report is available in Balacheff (1988, pp. 465–562). I will only focus here on the outcomes relevant to my present purpose. They are related to the observations which were made in two different classrooms.

The First Experiment

The first experiment has been carried out with students of the eighth grade (13 to 14 years old). The teacher was a member of the research team, which means that we were in a good position to assert that the project was well known to her. The observations lasted for two sessions of 1:30 hours. After the first one we felt really happy with what had happened. After the second phase the feeling of some difficulties arose ... Beyond these feelings only the close analysis of the data gathered, led us to discover the existence of the parameters which had played a critical role in the teacher's decisions, and thus in the students' behaviors:

- First, a constraint of time, which made the teacher intervene in order to ensure that the whole process will keep within the limits imposed by the general school time-table in which the experiment took place.
- Second, the teacher's willingness to guarantee an acceptable end in her own eyes.

There was a huge tension between this willingness and the willingness of not breaking the contract of "non intervention". This tension is the indication of what we would like to call in the future: *the teacher's epistemological responsibility.*

Because of these two constraints, the decisions of the teacher tended to oppose the devolution of the problem. In particular, to guarantee that the problem solving phase will not be too long, the teacher invited students to propose a solution as soon as she thought that what they had got was mature enough, but with no information about the real feeling of these students. Also, some teacher's interventions aimed at calling students' attention to the word "any" (in the sentence "any triangle"). In doing so she did not think that it was a mathematical intervention, insofar as she thought that it was only due to the students' lack of care. But all these interventions led students towards a feeling of dependence and the idea of a possible responsibility of the teacher for the validity of their answer.

A significant phenomenon is that the teacher (as well as the observers) did not realize what a continuous contact she kept with the students, making about one intervention every minute over a 80 minutes period. The content of these interventions could have been light, as: "Are you O.K.?", or more important such as: "Are you sure you have carefully read the statement of the task?". Altogether we have counted, within these interventions, 129 different items. We see this phenomenon as an indicator of the intensity of the relationships between the teacher and the students in a situation that we thought to be quasi-isolated from the teacher before we did a close analysis of the records.

The same constraints were an obstacle to the functioning of the second phase. After a first exchange of criticisms by the teams' representatives, the teacher intervened because she thought that nothing positive would come out of the engaged process – at least within the time available. The teacher then tried to facilitate the progress of discovery of a solution, calling explicitly for ideas and suggesting that the students should start from them to go further. Actually it was quite clear from her attitudes that not all the ideas were of the same value:

> "People have said to me that perpendicular bisectors seem interesting, so try to work more on perpendicular bisectors ... so using perpendicular bisectors, try to resolve the problem stated ... We will rewrite a message, if you want to redo the work you did yesterday ... but not starting off quite like yesterday, because you've already brought out some ideas. OK, we're going to try to share the work. Well ... There are three ideas..."[11] (Balacheff 1988, p. 518).

The students' behaviors were deeply transformed by these interventions, they got confused and they were no longer committed to any real discovery of a solution.

The teacher thought that she had kept the spirit of the sequence, the basic frame being: search for a solution, criticisms, new ideas and suggestions to go ahead. But only the superficial aspects of the intended sequence were still there, their meaning for the students were fundamentally changed. They did not enter into a true mathematical activity, as expected, but just a new school game which was not so different, beyond the new and exciting social setting, from the ordinary one. As the teacher acknowledged (off the record): "Now they are not searching any more, they're ready to accept just about my message ... although they had arguments against ... they didn't bring them out"[12] (*ibid.*, p. 527).

The Second Experiment

We learned a lot from this first experiment, and we thought that it would be worthwhile to make a second one. We decided to keep the same general framework, but to overcome the obstacles we came to be faced with, we chosen the three following modifications:

(i) To observe a tenth grade classroom in order to be sure that no mathematical difficulty will disturb the phenomenon we wanted to observe. Also, at this level students have already been introduced to mathematical proof. The situation could be an opportunity to enlighten its power as a means for proving ...

(ii) To open the time, that means that we decide to leave open when the end of the experiment will occur. We thought that three or four sequences of about one hour will be sufficient.

(iii) To ask the teacher not to intervene, as strictly as possible, during the first phase (the initial problem-solving phase), and then to act just as a chairperson and as the collective memory[13] of the class during the second phase (the debate).

The first phase[14] did not present anything specific. The teacher did not intervene at all, leaving students free to decide whether they have a solution to propose. Four teams among the five reached a solution, the fifth one which was clearly close to surrender, finally proposed a "contribution" to the collective effort, as a response to the teacher demand.

During the second phase also she followed the specifications we decided together. Then ...

More than a scientific debate, that is, proposing proofs of counterexamples, the data show that students entered a discussion with some mathematical content in it, but which mainly consisted of an exchange of arguments *pro et contra* not necessarily connected one to another. They argue about the different proposed solutions, but they do not prove mathematically.

The situation for communication has really been taken into account as such by the students, as their remarks on the proposed messages show. The main criticisms are related to the fact that this message must be understandable and usable by its recipients:

"It must be understandable for everyone"[15] (C 920).
"We have to put ourselves in the position of someone who doesn't know"[16] (E 559).

But the problem of the validity of the proposed solution is not really considered. In that sense we can say that the situation does not become a situation for validation. For a clear distinction between "arguing" and "proving" in mathematics[17], we refer to the distinction as formulated by Moeschler:

"Argumentation discourse is not one which is brought to bear on proofs, nor a discourse which functions according to the principles of logical deduction. In other words, arguing does not aim to demonstrate the truth of an assertion, nor to show the logical validity of reasoning [...] Argumentative discourse – and this is already an important hypothesis – always stands in relation to a 'counter discourse', whether real or virtual. Argumentation, so understood, is thus indissociable from polemic." (Moeschler, 1985, pp. 46–47, our free translation)

In that sense, what we have observed is first of all an exchange of arguments about the simplicity of the solution:

"I find it's too simple ... I would say that they don't prove mathematically how to find the perimeter of a triangle, they only say ... they draw the lines. Anyone could do that."[18] (B 224).

... or of its complexity:

"OK, we're all agreed on the fact ... That we reject the message ... OK, rejected: Firstly, it's too complicated ... too complicated, not clear enough"[19] (E 434).

The context of a communication with other students has favoured the feeling of the relevance of critics in that register. But what leads us to suggest that this debate is more an argumentation than a scientific debate, in the Moeschler sense, is the frequent lack of logical relationships between arguments. Even more, some students can move in the same argumentation from one position to another which is completely contradictory:

"We're all agreed on refusing ... well, because firstly the message is too complicated and next, it's no longer a message, it is the whole problem. They do the problem, they calculate, they already do the calculations"[20] (D 458)
"Anyway, it's easy, since they already put everything there for us."[21] (D 592).

These arguments can have nothing to do with mathematics as such, or even with what is required by the situation:

"For a problem like that, it's too simple ... When you see that message, and then when you see this message [message B] you can guess that maths is a complicated thingie."[22] (C 230)

... and it could be the same for the objections opposed to an argument, like in the following example:

This argument is opposed to the argument of team B against the too easy message A (cf. above, B 224): "if he said it's too easy, that's 'cause he could do it, so he has not to write that he'd refused"[23] (E 275).

Finally, the involvement of some of the teams in the game, I mean the fact that they are eager to win, had favoured the appearance of polemics: The strongest opponents to the "too complex" message are the authors of the "too simple", and inversely.

After a first period of debate the messages have been accepted, provided that some modifications are to be made, but their validity has not been really discussed. So, the teacher proposed a new triangle, in order to challenge the messages. This triangle was such that the wrong solutions will obviously fail. The debate following this checking phase, shows how strongly students are more involved in an argumentation than in a scientific debate:

"Just now it was double"[24] (D 1465)
"We redid the message for them with the other piece of paper; it's not our fault if they gave us a smaller one"[25] (D 1492)
"Yes, but it's an ordinary triangle ... it's got to work in all cases. Here it doesn't work"[26] (E 1493)
"That shows that it doesn't work with any triangle, that's all"[27] (D 1501)
"Just now it worked alright there, why didn't it work here ... it might be a stroke of luck"[28] (D 1550)

Finally, one solution being accepted as the solution of the class, the teacher asked students whether they were sure of that solution. They answer: "Yes because we have done it in a lot of cases." So, it is not even sufficient to directly address the question of the validity. Note that when later on the teacher asked the students about a possible mathematical proof of their solution, they gave one showing that technically it was in their capability.

DISCUSSION

Efficiency Versus Rigor

Even if we are able to set up a situation whose characteristics promote content-specific students' interaction, we cannot take for granted that they will engage in a "mathematical debate", and finally that they will produce a mathematical proof.

A peculiarity of mathematics is the kind of knowledge it aims at producing. Its main concern is with concepts specific to its internal development. There is evidence that the Egyptians used intellectual tools in practical situations for which we have now mathematical descriptions, but the birth of mathematical proof is essentially the result of the willingness of some philosophers to reject mere observation and pragmatism, to break off perception (*le monde sensible*), to base knowledge and truth on Reason. That actually is an evolution, or a revolution, of mathematics as a tool – encapsulated in practice – towards mathematics as an object by itself made available – through a constructive process – for the purpose of explicit considerations, and as a consequence a change of focus from "efficiency" towards "rigor".

It is a rupture of the same kind which happens between "practical geometry" (where students draw and observe) and "deductive geometry" (where students have to establish theorems deductively). Also in numerical activities, like the one reported by Lampert (1988), the same rupture happens when students have no longer to find some pattern out of the observation of numbers, but when they have to establish numerical properties in their "full" generality (using letters and elementary algebra).

We have here to realize that most of the time students do not act as *theoreticians* but as *practical persons*. Their task is to give a solution to the problem the teacher has given to them, a solution that will be acceptable with respect to the classroom situation. In such a context the most important thing is to be effective. The problem of the practical person is *to be efficient* not *to be rigorous*. It is to produce a solution not to produce knowledge. Thus the problem solver does not feel the need to call for more logic than is necessary for practice.

That means that beyond the social characteristics of the teaching situation, we must analyse the nature of the target it aims at. If students see the target as "doing", more than "knowing", then their debate will focus more on efficiency and reliability, than on rigor and certainty. Thus again argumentative behaviors could be viewed as being more "economic" than proving mathematically, while providing students with a satisfactory feeling about the fact that they have completed the task.

Social Interaction Revisited

Social interaction, while solving a problem, can favour the appearance of students' proving processes. Insofar as students are committed to finding a common solution to a given problem, they have to come to an agreement on the acceptable ways to justify and to explain their choices. But what we have shown is that proving processes are not the only processes likely to appear in such social situations, and that in some circumstances they could even be almost completely replaced by other types of interactional behaviors. Our point is that in some circumstances social interaction might become an obstacle, when students are eager to succeed, or when they are not able to coordinate their different points of view, or when they are not able to overcome their conflict on a scientific basis.[29] In particular these situations can favour naive empiricism, or they can justify the use of crucial experiments in order to obtain an agreement instead of proofs at a higher level (Balacheff, 1988).

Perhaps some people might suggest that a better experiment in didactical engineering could allow us to overcome these difficulties; indeed much progress can be made in this direction and more research is needed. But we would like to suggest that "argumentative behaviors" (i) are always potentially present in human interaction, (ii) that they are genuine epistemological obstacles[30] to the learning of mathematical proof. By "argumentative behaviors" we mean behaviors by which somebody tries to obtain from somebody else the agreement on the validity of a given assertion, by means of various arguments or representations (Oléron, 1984). In that sense, argumentation is likely to appear in any social interaction aiming at establishing the truth or falsehood of something. But we do consider that argumentation and mathematical proof are not of the same nature: The aim of argumentation is to obtain the agreement of the partner in the interaction, but not in the first place to establish the truth of some statement. As a social behavior it is an open process, in other words it allows the use of any kind of means; whereas, for mathematical proofs, we have to fit the requirement for the use of some knowledge taken from a common body of knowledge on which people (mathematicians) agree. As outcomes of argumentation, problems' solutions are proposed but nothing is ever definitive (Perelman, 1970, p. 41).

As far as students are concerned, we have observed that argumentative behaviors play a major role, pushing to the background other behaviors like the one we were aiming at. Clearly enough, that could be explained by the fact that such behaviors pertain to the genesis of child's development in logic: Very early, children experience the efficiency of argumentation in social interactions with other children, or with adults (in particular with their parents). Therefore it is quite natural that these behaviors appear first when what is in debate is the validity of some production, even a mathematical one.

So, what might be questioned is perhaps not so much the students' rationality as a whole, but the relationships between the rationale of their behaviors and the characteristics of the situation in which they are involved. Not surprisingly, students refer first to the kind of interaction they are already familiar with. Argumentation has its own domain of validity and of operationality, as all of us know.

So, in order to teach mathematical proof successfully, the major problem appears to be that of how to *negotiate* the acceptance by the students of new rules, but not necessarily to require that they reject argumentation insofar as it is perhaps well adapted to other contexts. Mathematical proof should be learned "against" argumentation, bringing students to the awareness of the specificity of mathematical proof and of its efficiency to solve the kind of problem we have to solve in mathematics.

Negotiation is the key process here, for the following reasons:

- First, because the teaching situation cannot be delivered in an "open" way to the students otherwise many of them will not understand the point and they will get lost. The following quotation from Cooney makes it clear:

> "Maybe not all of them but at least some of them felt 'I am not going to participate in this class because you [referring to the teacher] are just wasting my time'. It is so ironic because if I was doing the type of thing they wanted to do, they would be turning around in their seats and talking. So it's a no-win situation." (Cooney, 1985, p. 332).

- Second, because of the rules to be followed, the true aim of the teacher cannot be stated explicitly. If the rules for the interaction are explicitly stated, then some students will try to escape them or to discuss them just as many people do with law. Also because interacting mathematically might then become "mastering a few clever techniques" which may turn into objects to be taught, just as teaching "problem solving" has often become teaching quasi-algorithmic procedures (Schoenfeld, 1988).

The solution is somewhere else, in the study and in the better understanding of the phenomena related to the didactical contract, the condition of its negotiation, which is almost essentially implicit, and the nature of its outcomes: The devolution of the learning responsibility to the students. We cannot expect "ready-to-wear" teaching situations, but it is reasonable to think that the development of research will make available some knowledge which will enable teachers to face the difficult didactical problem of the management of the life of this original society: The mathematics classroom.

NOTES

[1] Programmes fixés par l'arrêté ministériel du 14 november 1985.
[2] Quoted by Braconne 1987, p. 10. Free translation of the French text: "Développer l'esprit logique de l'élève en l'amenant à structurer ses raisonnements [...] Il convient donc d'exiger que l'élève appuie la solution des problèmes sur un raisonnement logique et structuré".
[3] In a precise experimental situation (Balacheff, 1982).
[4] This word stands here for the French: "démonstration", used by Braconne in her text.
[5] The interviewees were 13 French mathematics teachers.
[6] Manin quoted by Hanna (1983).
[7] The essay of I. Lakatos (1976) on the dialectic of proofs and refutations gives a good insight into this historical process.
[8] By "common", we mean here a solution supported by the whole classroom, or smaller groups of students as it is usually the case.
[9] The case study reported here has been made possible because of the close relationships established between academics and teachers within a research group of the *IREM de Lyon.* It is a small part of a four years project which had allowed us to collect a large amount of data. The complete report is available from the *IREM de Lyon, Université Claude Bernard, Lyon.*
[10] Devolution" "A delegating of authority or duties to a subordinate or substitute" (The American Heritage Dictionary of the English Language, 1979).
[11] Since that has been translated freely from a protocol, we give here the original text: "Il y en a qui m'ont dit que les médiatrices ça semble intéressant, donc essayer de retravailler sur les médiatrices ... en utilisant donc les médiatrices essayez de résoudre le problème qui vous est posé ... on va refaire un message, si vous voulez refaire le même travail que vous avez fait hier ... mais en partant pas tout à fait comme hier, car vous avez déjà des idées qui sont sorties. Alors on va essayer de répartir le travail, je ne sais pas là moi ... il y a trois idées..."
[12] Our free translation, from the French: "Là, ils ne cherchent plus, ils sont prêts à accepter n'importe quel message ... alors qu'ils avaient des arguments contre ... ils ne les ressortent pas."
[13] To be the "memory" of the class means to take a record of what is said, in particular by writing students' decisions on the blackboard.
[14] This phase took about one hour.
[15] Our free translation, from the French: "Il faut qu'il soit compréhensible pour tout le monde".
[16] Our free translation, from the French: "Faut se mettre à la place de quelqu'un qui sait pas ..."
[17] We do not refer necessarily to formal proof, or mathematical proof in the classical

sense.

[18] Our free translation, from the French: "Je trouve que c'est trop simple ... je veux dire, ils démontrent pas comment trouver le périmètre d'un triangle, ils disent juste ... ils tracent des droites. Tout le monde peut le faire."

[19] Our free translation, from the French: "Alors, nous on est tous d'accord sur le fait ... qu'on refuse le message ... Alors refusé: premièrement il est trop compliqué ... trop compliqué, pas assez clair."

[20] Our free translation, from the French: "On est tous d'accord pour refuser ... alors, parce que d'abord le message est trop compliqué et puis c'est plus un message, c'est le problème en entier. Ils font le problème ... ils calculent, ils font déjà tous les calculs."

[21] Our free translation, from the French: "C'est facile quand même, puisqu'ils nous mettent déjà tout."

[22] Our free translation, from the French: "Pour un problème comme ça c'est trop simple ... quant on voit ce message, pis quant on voit celui-ci [message B] on pourra deviner que les maths c'est des machins compliqués."

[23] Our free translation, from the French: "S'il dit que c'est trop facile c'est qu'il a réussi à le faire, donc il n'a pas à marquer qu'il a refusé."

[24] Our free translation, from the French: "Tout à l'heure il était réalisable."

[25] Our free translation, from the French: "Nous on l'a refait le message avec l'autre feuille, c'est pas notre faute s'ils nous en donnent une plus petite."

[26] Our free translation, from the French: "Oui, mais c'est un triangle quelconque ... Y faut que ça marche dans tous les cas. Ici ça marche pas."

[27] Our free translation, from the French: "Ça prouve que ça marche pas avec n'importe quel triangle, c'est tout."

[28] Our free translation, from the French: "Tout à l'heure il a bien marché là, pourquoi il n'a pas marché ici ... ce serait un coup du hasard."

[29] I mean, content specific.

[30] The notion of "epistemological obstacle" has been coined by Bachelard (1938), and then pushed on the forefront of the didactical scene by Brousseau (1983). It refers to a genuine piece of knowledge which resists to the construction of the new one, but such that the overcoming of this resistance is part of a full understanding of the new knowledge.

REFERENCES

Arsac, G. and Mante, M.: 1983, 'De "problèmes ouverts" dans nos classes du premier cycle, *Petix x* 2, 5–33.

Bachelard, G.: 1938, *La formation de l'esprit scientifique*, Vrin, 1975, Paris.

Balacheff, N.: 1982, 'Preuve et démonstration au collège', *Recherches en didactique des mathématiques* 3(3), 261–304.

Balacheff, N.: 1988, 'Une étude des processus de preuve en mathématiques chez des élèves de collège', Thèse d'état, Université Joseph Fourier, Grenoble.

Balacheff, N.: 1988. 'Aspects of Proof in Pupils' Practice of School Mathematics', in Pimm, D. (ed.), *Mathematics, Teachers and Children*, Hodder and Stoughton, London, pp. 216–235.

Bell, A.: 1976, 'The Learning of General Mathematical Strategies', Ph.D. University of Nottingham.

Bishop, A.J. 1988, *Mathematical Enculturation: A Cultural Perspective on Mathematics Education*, Kluwer Academic Publishers, Dordrecht.

Bourdieu, P.: 1980, *Le sens pratique*, Edition de Minuit, Paris.

Braconne, A.: 1987, *Compréhension de la démonstration en géométrie chez les profes-*

seurs et les élèves au secondaire, Maitrise ès Art, Université de Laval, Québec.

Brousseau, G.: 1983, 'Les obstacles épistémologiques et les problèmes en mathématiques', *Recherches en didactique des mathématiques* **4**(2), 164–198.

Brousseau, G.: 1986, 'Basic Theory and Methods in the Didactics of Mathematics', in Verstapen, P.F.L. (ed.), *Second Conference on Systematic Co-operation Between Theory and Practice in Mathematics Education*, Instituut voor Leerplanontwikkeling, Enschede, pp. 109–161.

Capponi, M.T. *et al.*: 1986, *Apprentissage du raisonnement*, IREM de Grenoble, Grenoble.

Cobb, P.: 1986, 'Contexts, Goals, Beliefs, and Learning Mathematics', *For the Learning of Mathematics* **6**(2), 2–9.

Cooney, T.J.: 1985, 'A Beginning Teacher's View of Problem Solving', *Journal for Research in Mathematics Education* **16**, 324–336.

Fawcett, H.P.: 1938, *The Nature of Proof.* 1938 NCTM Year Book. Columbia University Teachers College, New York.

Galbraith, P.L.: 1979, *Pupils Proving*, Shell Center for Mathematics Education, University of Nottingham.

Gaud, D. and Guichard, J.P.: 1984, 'Apprentissage de la démonstration', *Petit x* **4**, 5–25.

Hanna, G.: 1983, *Rigorous Proof in Mathematics Education*, Curriculum Series, 48, The Ontario Institute for Studies in Education, Toronto.

Inhelder, B. and Caprona, D.: 1985, 'Constructivisme et création de nouveauté', *Archives de psychologie* **53**(204), 7–18.

Lakatos, I.: 1976, *Proofs and Refutations*, Cambridge University Press, Cambridge.

Lampert, M.: 1988, 'The Teacher's Role in Reinventing the Meaning of Mathematical Knowing in the Classroom, in *Proceedings of the PME-NA*, pp. 433–480.

Martin, G. and Harel, G.: 1989, 'Proof Frames of Preservice Elementary Teachers', *Journal for Research in Mathematics Education* **20**(1), 41–51.

Moeschler, J.: 1985, *Argumentation et conversation*, Hatier-Credif, Paris.

NCTM: 1989, *Curriculum and Evaluation Standard for School Mathematics*, NCTM, Reston.

Oléron, P.: 1984, *L'argumentation*, PUF, Paris.

Perelman, C.: 1970, *Le champ de l'argumentation*, Presses Universitaires, Bruxelles.

Popper, K.R.: 1979, *Objective Knowledge*, Revised Edition, Oxford University Press, Oxford.

Schoenfeld, A.H.: 1988, 'When Good Teaching Leads to Bad Results: The Disaster of Well Taught Mathematics Courses', *Educational Psychologist* **23** (2).

Schoenfeld, A.H.: 1990, 'On Mathematics as Sense-Making: An Informal Attack on the Unfortunate Divorce of Formal and Informal Mathematics', in Perkins, D.N., Segal, J. and Voss, J. (eds.), *Informal Reasoning in Education*, Erlbaum, Hillsdale, pp. 311–343.

Senk, S.: 1985, 'How Well Do Students Write Geometry Proofs?, *Mathematics Teacher*, **74**, 19–26.

Usiskin, Z.: 1982, *Van Hiele Levels and Achievement in Secondary School Geometry*, CDASSG Project, SE 038 813.

ENTR'ACTE 9

We stay in the 'social' arena of the classroom for this final chapter, while we approach the problems of development in another area of mathematical knowledge. Where earlier chapters have referred to specific mathematical content, this next chapter, by Bishop, focusses the reader's attention on the issues surrounding the teaching of values in mathematics education. This is a rarely examined aspect of mathematics teaching and indeed many people would probably assume that mathematical knowledge is somehow value-free. We could see, however, in the previous chapter how the pupils were not just dealing with a mathematical topic; they were also grappling with what Morris Kline calls "the spirit of rationality". There is more to a mathematical education than the syllabus content words like 'fractions, decimals, ratios, etc.' and Bishop argues that mathematical values are implicity present in all mathematics teaching, and that they cause educational problems precisely because they remain implicit. He presents his interpretation of the most significant mathematical values, and analyses firstly how they impinge on the teaching/learning situation, and secondly how mathematics teachers could avoid the negative consequences of their implicitness by finding strategies to make their contribution more explicit.

ALAN J. BISHOP

MATHEMATICAL VALUES IN THE TEACHING PROCESS

1. INTRODUCTION

The fact that mathematics is a part of every pupil's school curriculum is rarely questions – it is indeed difficult to find a country where mathematics is not a significant part of the curriculum. Yet we also know that in most countries many pupils continue to experience a great deal of difficulty with the subject. Mathematics educators the world over are continually searching for ways to improve this situation. "Maybe the pupils don't enjoy it enough" it is said, so there are strategies to make mathematics more fun. Or "Maybe they don't see the point of it enough", so the search is on for more 'relevance'. Research is continually directed at this problem, the possibilities and probabilities are weighed and assessed, and the debate goes on.

A new perspective on this debate is the aspect of the *values* associated with the mathematics the pupils are being taught. It is perhaps rather surprising that it is only in recent years that attention has begun to be paid to this issue in mathematics education, since values associated with other subjects on the school curriculum are often debated, and the phenomenon of the 'hidden curriculum' (Apple, 1980) is well known.

However, from various studies we know that pupils *do* develop certain attitudes towards mathematics, that adults *do* have views about the relevance and importance of mathematics, that choices *are* made by pupils when comparing mathematics with other subjects, and that teachers *do* have individual ideas about the importance and aims of studying mathematics. These value-related aspects of mathematics *are* learnt, and they are probably learnt implicitly and unintentionally. They are therefore likely to be influential to an extent not always recognised and thus may not necessarily be beneficial to pupils' mathematical education.

For example, Buxton (1981) summarises his research on adults' fear of mathematics, and says that mathematics is largely seen as:

1. Fixed, immutable, external, intractable and uncreative.
2. Abstract and unrelated to reality.
3. A mystique accessible to few.
4. A collection of rules and facts to be remembered.
5. An affront to common sense in some of the things it asserts.
6. A time test.
7. An area in which judgements not only on one's intellect but on one's

Alan J. Bishop et al. (eds.), Mathematical Knowledge: Its Growth Through Teaching, 195–214.

personal worth will be made.
8. Concerned largely with computation.

He adds "Some may seem to contradict others, but that is no bar to their being held at the same time, by the same person. Perhaps it is only within mathematics that one should expect such consistency!" (p. 115).

However, far from being considered important, many teachers and others in education seem to believe that mathematics is value-free knowledge, and therefore it is rare to find a mathematics classroom in which values play an *intentionally* significant role, or a mathematics textbook in which there is any serious discussion *about* mathematics, or a school examination in which questions *about* mathematics are asked. Mathematics in school still seems to be, in the majority of cases, a performance subject, where skills are emphasised, and such values which are being developed throughout a pupil's mathematical education are destined to remain hidden, unknown and unanalysed by an education which doesn't question and doesn't provoke.

It is also rather rare to find any mathematics educator prepared to accord values a major significance in a pupil's mathematical education – hence this chapter (two notable exceptions are Frankenstein, 1989 and Wilson, 1986). So, I intend firstly to present a case not just for their recognition in the mathematics curriculum, but also for their *explicit* consideration in the mathematics classroom. Also I intend to analyse the mathematics teaching process in relation to values, in order to show how it is possible to change the present emphasis, which in many cases appears to be accidental and largely negative, to one which will enable more pupils to make progress in understanding mathematical knowledge. I am therefore concerned with the didactical genesis and growth of values in mathematics, and about their influence on pupils' conceptions of mathematical knowledge.

2. VALUES AND MATHEMATICAL KNOWLEDGE

The background to the analysis in this chapter is given in Bishop (1988) and it rests centrally on the notion that mathematics is part of everyone's cultural heritage. Mathematical activities exist in all cultures but they seem to be particularly significant in industrialised societies. Mathematical knowledge influences almost every aspect of today's technologically-based societies and in my view, it therefore needs to be better understood and thoroughly discussed. The reason mathematics is a part of everyone's curriculum is, I would argue, because it is a part of everyone's culture.

Lancy (1983) has demonstrated that pupils learn any aspect of their culture's knowledge at three levels – firstly, their cognitive structures are developing, while they are learning anything; secondly, they learn the overt specifics of cultural knowledge, the language and other symbolisations, the way their cultural group categorises their world; thirdly, the pupil learns 'about' knowledge, the ideas and the principles behind the symbols used by their culture, and they learn about what values their cultural group associates with this knowledge. Thus mathematical knowledge can influence at all three levels, and like any other form of cultural knowledge, mathematical knowledge has values associated with it.

To recognise mathematics as a cultural product is also to confirm that it is a human product. As such it exists because people have valued the activities concerned with it, and it remains significant because people continue to value it. Nothing which exists can be said to have any more intrinsic value than any other thing, but human creations exist and continue to exist because humans have chosen to create them and because humans, either individually or socially, have continued to value them.

But people value mathematics for different reasons. A parent might think mathematics is important for their child to study, so that they can get a good job later on. The child however might choose to study mathematics for the enjoyment she gets out of exploring abstract ideas. The teacher might like to teach mathematics for the rigorous training of the mind that it is assumed to develop in the young. How should we handle such differences?

In the educational context we particularly recognise the existence of the values of mathematics whenever we attempt to answer the question "Why should we teach mathematics to our pupils?" Answers to this question vary historically and culturally, but here are three rather diverse examples to demonstrate and exemplify the context to which I am referring.

In the UK, the influential Dainton Committee of 1968 listed the following reasons for teaching mathematics, all of which imply a recognition of values:

1. Mathematics as a means of communicating quantifiable ideas.
2. Mathematics as a training for discipline of thought and for logical reasoning.
3. Mathematics as a tool in activities arising from the developing needs of engineering, technology, science, organisation, economics, sociology, etc.
4. Mathematics as a study in itself, where development of new techniques and concepts can have economic consequences akin to those flowing from scientific research and development.

As a contrast Swadener and Soedjadi (1988) referred to the following important aspects in their discussion of values-related ideas of mathematics in Indonesia:

A. Universe

In every mathematical problem it is important to understand the existence of a *universe* for the problem. The solution to a given mathematical problem will likely be different within different universes.

B. Agreement/Conventions

Mathematics is replete with agreements about concepts, facts, and operations. Agreements in mathematics depend on the *freedom* to define mathematical "objects" as necessary to accomplish desired goals.

C. Contradiction

Contradictory results are not permitted in mathematics. If contradiction is permitted in a mathematical structure, the structure collapses and is therefore unproductive.

D. Transformation

In mathematics there are many formulas which translate one set of conditions in one universe into another set of conditions in another universe. Several transformation formulas may be used to change one quantity into another or to translate a relation in one universe into another form in another universe.

E. Analogy

In mathematics the teacher can pay special attention to analogies, i.e. similarities between sets of circumstances with respect to both form and procedure.

Thirdly, one report of the Second World Conference on Muslim Education, held in Islamabad, Pakistan in 1980, includes this description of the purposes of mathematics teaching, at the primary level:

> By the time a student finishes the primary stage he should be able to handle algebraic symbols and geometric figures with confidence in addition to arithmetic numbers. The objective is to make pupils implicitly able to formulate and understand abstractions and be steeped in the area of symbols. It is good training for the mind so that they may move from the concrete to the abstract, from sense experience to ideation and from matter-of-factness to symbolization. It makes them prepare for a much better understanding of how the Universe, which appears to be concrete and matter of fact, is actually *ayatullah*: signs of God – a symbol of reality.

The problem is that aims statements like these above seem so haphazard and unstructured. All these examples do is to raise questions of what *are* these values, how can we recognise them, and how should we conceptualise them?

Firstly we need to recognise the *process* of valuing, which is about expressing preference or choice. For example, we choose to include mathematics in the school curriculum, rather than say, gardening, or numismatics, because of certain qualities which it possesses and which are valued by society – for example, as they are described above by the Dainton Committee. Mathematics has certain qualities or attributes as knowledge, which makes it *more* important than gardening or numismatics for all pupils to study. Swadena and Soedjadi, and the Muslim Education conference choose other aspects of mathematics as being the important qualities to focus on. We value mathematics because of perceived qualities.

Moreover, we also value these mathematical qualities at different levels of social grouping, a fact which helps to explain the rather disparate nature of aims statements, for example. At the *societal* level we can recognise the particular valuations put on mathematics by society. In Western industrialised societies, for example, mathematical qualifications are highly valued and lead to higher status jobs. Mathematical questions are put in popular mental tests because it is assumed that they will test a higher level of intelligence. Mathematics also impinges on almost every aspect of industrial and commercial life through its power and control.

At the *institutional* level we can see how the valuations of mathematical qualities affect within-school curricular and organisational practices. Mathematics is rarely integrated with any other subject, and is usually held in high esteem. It is a central part of any school curriculum, and success in it will be necessary for progressing to further studies in a wide range of fields in higher education. Schools, of course, do mediate society's interpretations and beliefs about mathematics, but they can also inject their own valuations.

In the same way teachers, at the *pedagogical* level, can prefer to, and *do*, value some qualities of mathematics rather than others. These preferences emerge in their teaching, resulting in the creation of rather different mathematical knowledge environments for the pupils in their care.

As a result of this, values at the level of the *individual* pupil will differ markedly, and we see the results emerging in the various attitude and value statements which pupils offer. There are of course the "maths-rejectors", those pupils who choose not to value any of the qualities of maths, as realised in their classrooms, although whether they would reject mathematics quite so demonstrably if the schools and teachers realised other qualities of mathematics, is an open question.

One thing we can see from this analysis is that the values felt in the classroom are derivative – they derive from qualities and values expressed within both schools and society at large. Indeed all these values must derive in some way from qualities, beliefs and values about the mathematics *itself*; that is, mathematics at the level of *cultural* knowledge.

It is therefore that level to which we must now turn.

3. VALUES AND THE QUALITIES OF WESTERN MATHEMATICAL KNOWLEDGE

The recognition that mathematics is a cultural product does not imply that all cultures have necessarily developed the *same* mathematical knowledge. All cultures develop language, yet languages are not the same around the world. The same is true for religion, morals, customs, kinship systems and games. All cultures develop these but the products and symbolisation in one culture are not necessarily the same as those in another culture. There are similarities but there are also differences.

Mathematics can now be considered in the same way. The historical literature makes reference to Indian mathematics, Chinese mathematics and Islamic or Arabic mathematics, and more recent anthropological and cross-cultural data also point to the fact that all cultures seem to develop mathematical ideas but not necessarily the same mathematical ideas. There are recognisable similarities, but there are also differences (see e.g. Zaslavsky, 1973; Pinxten, 1983; Gerdes, 1988). We now therefore face a difficulty in talking about mathematics, because we need to ask questions like "Whose mathematics is it?" and "Which mathematics are we talking about?"

So, let me first of all make it clear that the mathematics to which I refer, as do all the other authors in this book, and whose values I will explore, is what is now called "Western Mathematics". I say this because it is likely that other cultural traditions will have developed different values in their mathematics. There are certainly hints in the literature about such value differences. For example, Lawler (1982) in his powerful book on Sacred Geometry says: "We have seen how in India the adoption of zero was associated with a doctrine which negated the reality of the material world" whereas "The less mystical and more practical orientation of the Arab mentality saw in these (Indian) symbols a practical device for facilitating calculation and recording large numbers". Again "The 'western' rationalistic mentality negated the ancient and revered spiritual concept of Unity, for with the adoption of zero, Unity loses its first position and becomes merely a quantity among other quantities" (pp. 18, 19).

As another example, Ronan (1981) quoting Needham's voluminous data on Chinese mathematics and science says: "There was (in Chinese mathematics) an absence of the idea of rigorous proof, possibly as a result of the mental outlook which avoided the development of formal logic in China and which allowed associative or organic thinking to dominate" and "Mathematics in China was therefore utilitarian, its social origins bound up with the bureaucratic government system, and devoted to the problems the ruling officials had to solve" (pp. 62, 63).

As another point of reference, Western mathematics is the subject of Kline's (1972) book "Mathematics in Western culture", and even though there is now plenty of dispute about the history and development of Western mathematics (e.g. Joseph, 1987), the relationship with Western culture in recent centuries is clear. "Western mathematics" is actually rather a misnomer, since not only have many cultural groups contributed to its development, but there are many practising mathematicians in the world who would reasonably object to being thought of as part of the Western cultural tradition. Another label, Mainstream mathematics, is perhaps more acceptable although one could certainly object to its exclusive character which would marginalise all other mathematical developments such as ethnomathematics (d'Ambrosio, 1985), and folk mathematics (Mellin-Olsen, 1987). Furthermore I believe that in order to consider the values aspect of mathematics seriously, it is very necessary to attach the label 'Western', rather than one which is less culturally located, such as 'Mainstream', since it was the 'Westernising' of mathematics within Western Europe which crystallised the values now associated with it. This is also the reason why at this level mathematics shares its values with other knowledge fields like science.

In my analysis I have been uncovering the values and qualities which appear to have shaped Western Mathematical knowledge and which are discernible through texts and socio-historical accounts of mathematical developments. My evidence is clearly not first-hand and the analysis is always open to challenge. However I believe it is important for mathematics educators to try to clarify this area and I therefore hope that my attempts will lead to better ones. For my work I have leaned heavily on the conceptual structure offered by White (1959) and as a result, I have identified six clusters of values associated with certain qualities of Western Mathematical knowledge which I believe need consideration, and which are related to each other as three sets of complementary pairs.

The first cluster of values concerns the ideas of *rationalism.* If one were searching for only one identifiable quality of Western mathematics it would be *this* cluster of values. It is rationalism which many historians agree has guaranteed the pre-eminence of mathematics within Western culture and thence throughout the world, not tradition, not status, not experience, not seniority. As Kline (1972) says, for example: "In its broadest aspect (Western) mathematics is a spirit, the spirit of rationality. It is this spirit that challenges, stimulates, invigorates, and drives human minds to exercise themselves to the fullest" (p. 26).

Levidow (1988) refers to this aspect by contrasting it with the Indian scientific values tradition:

"According to Dr. M.D. Srinivas, of the PPST Foundation, Madras, Western logic and mathematics have been defined as rigorous by virtue of formulating ideas in a purely symbolic language independent of content and context, such that a statement may be logically rigorous yet bear no relation to reality. By constrast, he argued, the Indian scientific tradition founded a technical language on ordinary Sanskrit by refining it to remove ambiguities yet without making it entirely abstract and formal. Within that language, one cannot formulate a logical statement without considering its concrete validity and social meaning" (p. 110).

Another important aspect of this value cluster is that it is not the concrete world which is logical, it is explanations which are. Rationalism requires abstractions, explanations, theories. Proof is a refined, logical explanation of a relationship between abstractions. To learn to abstract, or to explain, or to theorise, is a first pre-requisite for learning to rationalise. Western culture has very rich languages for logic – Gardner (1977) in his (English) tesets found over 800 different logical connectives. Even in everyday discourse, arguments and discussions are frequently bound by the rules of logic. Nor must we forget the aesthetic appeal of what is sometimes called 'cold' logic – the purity, the precision, the accuracy and the inevitability of a logical argument has much appeal in times of confusion and conflict!

As a complementary cluster of values to rationalism, I offer those related to "objectism". I label this quality as *objectism* to represent the fact that Western culture's world-view is dominated by the idea of 'objects'. Where rationalism is concerned with the logic of the relationship between ideas, objectism is about the genesis and phenomenology of those ideas. Properties and relationships are named as nouns, further properties and relationships between *these* ideas are explored and more relationships developed and objectivised through using further nouns. One of the ways Western mathematical knowledge has gained its power is through the activity of objectivising the abstractions from reality. Through its symbols (letters, numerals, figures) people have been enabled to deal with abstract entities *as if they were* objects. The quality of objectivist knowledge, as compared with processual knowledge is powerfully conveyed by this reliance on symbols and symbolic objects.

The early Pythagoreans were atomists, and Number was probably the first mathematical entity to be systematically developed. Later though came the search for axioms, particularly in Geometry, and of course axioms are the 'atoms' of argument from which complex propositions can be built. This cluster of values also contains the ideas of materialism and determinism which are firmly rooted in a preferred concrete and mechanistic characterisation of phenomena. Ever since Newton's time the 'mechanisation of the world picture' has been a powerful value-force in shaping Western mathematical developments (see e.g. Dijksterhuis, 1986).

Rationalism and objectism are arguably the twin driving qualities of Western mathematics. They, more than any other qualities, are what have enabled Western mathematics to "conquer" the world of mathematics, and to be valued above others.

Other qualities and values are however discernible, and they too need to be fully recognised because they play a significant part in people's valuing of mathematics. One important cluster of values is focussed on the idea of *control*. So much of the power of Western mathematics seems to come from the security and control that it offers. Mathematics, through science and technology has given Western culture strong feelings of security in knowledge. We feel that we are in control of phenomena when we can describe them mathematically, so much so that we can become very frustrated at natural or man-made disasters which we feel shouldn't have happened.

In another context, the inconsistency of a mathematical argument provides a strong motive for uncovering the error and getting the answer 'right'. The mathematical valuing of 'right' answers spills over into Western society which also looks (in vain, of course) for right answers to its problems. The more we see mathematical ideas as 'objective facts', obeying 'rules' and 'laws' apparently like other objects, and subject to checks and logical arguments, the more we are valuing the quality of control.

There is however a complementary cluster of Western mathematical values which is associated with *progress*. Whereas mathematical control is about security, stability and predictability, mathematical progress is about change and alternatives. The satisfactions of 'control' seem to urge us to seek yet more control. Firstly a mathematical method of solution for one problem is enabled, by the abstract nature of Western mathematics, to be generalised to other problems. The unknown can become known. Knowledge can develop. Secondly, mathematical progress is achieved by seeking out situations where generalistions do not hold. The search for the counter-example, the extreme case, and the logical inconsistencies leads to a new problem which must be mastered. Finally and more recently historically, there is the intentional and deliberate alteration of the 'givens' and assumptions in order to develop alternative structures. Alternativism is alive and well in Western Mathematics and is valued by being at the root of mathematical creativity in Western culture.

The third pair of value clusters concerns more specifically the relationships between people and ideas of Western mathematical knowledge. One cluster I call '*openness*' and this is to do with the quality that mathematical truths are open to examination by all. Provided of course that one has the necessary prerequisite knowledge, it is theoretically and practically possible for anyone and everyone to verify and examine the truth of any proposition. Opinions may be held by people, but Western mathematical knowledge is valued because it represents what people like to think of as 'facts'. They don't go out of date, they don't vary from one country to another – because of their abstract and generalised nature they are universal truths which can be verified by anyone in any country.

Proof grew from the desire for articulation and demonstration, so well practised by the early Greeks, and althouth the criteria for the acceptability of a proof may have changed, the quality of 'opening' the knowledge has remained as strong as ever. Western science also of course embodies this value and it has helped to feed Western culture's views of democracy and liberalism. The American 'Declaration of Independence' could be thought of as a supreme example of a mathematical value having a profound effect on a society! Like the clarity of a mathematical proof its importance lies in its generalised and objective existence, rather than in the mind of the human being who drafted it. Openness in mathematics is also, and perhaps necessarily, a dehumanised openness.

Therein, paradoxically it may seem, lies the root of the last mathematical quality of significance – which I call *mystery*. This is a complementary value cluster to that of openness, for despite the apparent openness, there is nevertheless a mysterious quality about mathematical ideas. Certainly everyone who has learnt Western mathematics knows this intuitively, whether it is negatively through the meaningless symbol-pushing which many pupils still unfortunately experience, or whether it is more positively in the tantalising discovery of an unexpected connection. (Did you know that if you multiply together the numbers of any Pythagorean triple the result is always a multiple of 60?) The basis of the mystery is again the highly abstract nature of Western mathematics – abstractions take one away from a context, and it is well known that decontextualised knowledge can become meaningless, which contributes to the mysterious quality. Of course mathematical ideas offer their own networks of context so it is always possible to develop meanings *within* Western mathematics. But where these notions and ideas 'come from' is a mystery to all.

There are two classic resolutions of this issue. One is to believe in the idea of a higher authority, a God, a deity, a spirit, which in some sense is the ultimate possessor and distributer of all knowledge. The predominant deity influencing Western cultural thought has been the God of Christianity, and the notion of 'God the Mathematician' was one which held sway in the sixteenth and seventeenth centuries. All that mere mortals can do is to discover the knowledge laid down by the Almighty.

The other resolution of this issue is to argue that ideas don't 'come from' anywhere. Western mathematical knowledge, it is argued, like all cultural knowledge, is invented and created by humans through social and interpersonal interaction. Mathematicians *create* mathematical knowledge, but this of course doesn't 'solve' the mystery not only because mathematicians themselves are not well known, but also because the 'person' of the mathematician is also believed to be largely irrelevant, as Wilder (1978) points out and as the value of 'openness' testifies. The fact that Pythagoras' name is associated with That Theorem doesn't tell us that he created it or postulated it or proved it, alone or before anyone else. Indeed it is well known that Chinese (Ronan, 1981) and African (Gerdes, 1988) mathematicians were aware of the relationship before the Greek era began. And so the mystery deepens – were there several independent discoveries of the same idea or did the information arise in one place and become transmitted to another? Van der Waerden (1983) and others for example argue strongly for the independent, multiple discovery hypothesis.

These then are what I suggest are the six principle clusters of beliefs and values which are associated with the various qualities of Western mathematics as we know it today. Far from being value-free, it is clear from its cultural history that Western Mathematics has been developing the values of rationalism, objectism, control, progress, openness and mystery, in human interactions with the Western cultural and social environment. Note that I am not claiming that these values are *only* associated with Western mathematics, nor that these values are superior to any others, although earlier educators certainly *have*, particularly in the Western European colonial era (see Bishop, 1990).

4. THE TEACHING PROCESS IN MATHEMATICS EDUCATION

Turning now to the teaching process itself there are some necessary assumptions to make clear before I can begin to theorise about this process in relation to Western mathematical values. (From now on I will drop the adjective Western unless I wish specifically to distinguish its effect. However the word 'mathematics' should also be read as 'Western mathematics' unless otherwise qualified.)

I believe that a mathematical education is provided by a teaching process which is characterised as follows. It is:

- interpersonal, between teacher and pupil, and between pupil and pupil. That is, I choose to ignore 'text-teaching' where the textbook *is* either the only source of information or else determines what the teacher says and does. Neither can in my view offer an education;
- concerned with mathematical concepts, meanings, techniques and values. That is, I do not consider 'training', where the mere performance of mathematical techniques is the only goal, to be capable of providing a mathematical education.
- formal, institutionalised, and accountable. That is, I am not considering *informal* teaching which happens casually and accidentally, nor *non-formal* teaching which, though not intentional is not part of an educational system.
- involves the whole person of the pupil. That is, 'impersonal learning', where the individuality of the learner, their feelings, ideas and views are ignored, cannot in my view provide an education. The full intellectual and emotional powers of the pupil need to be engaged if a mathematical education is to result.

In addition, I have already described my perspective on the cultural nature of mathematics. In essence then, the teaching process in mathematics is best understood as a *shaping* process whereby "concepts, meanings, techniques and values" are shaped according to certain criteria, namely mathematical criteria. It is therefore an intentional process of shaping certain ideas.

Intentional shaping in an interpersonal educational context needs however to be understood as a cooperative activity between the pupils and their social environment. Initially it seems that it is the teacher who is doing the shaping, but consider that it is "Mathematical concepts, meanings, processes and values" which are being shaped, and these *belong* to the pupil. They are derived by the pupil, constructed by the pupil, and of course, shaped by the pupil, in response to signals received from the teacher and from the rest of the social environment. The pupil is an adaptive and constructive organism.

A mathematical education needs both pupil and environment to play a strong role. The pupil is not to be thought of as a *receptive* vessel for mathematical knowledge. On the contrary the pupil is the person who must decontextualise and reconstruct the mathematical knowledge from the contextualised situation offered in the classroom. Although we can glibly talk of 'knowledge transmission', and although it is clearly possible to recognise similar knowledge existing in consecutive generations, it is the pupils in any one generation who are busy re-creating and reconstructing the mathematical knowledge of their parents' generation and who in their turn structure and recontextualise the mathematical knowledge into situations within which *their* children's generation can do their own de-contextualising and re-creation of the knowledge.

On the mathematics teacher's part there are the kinds of signalling activity which constrain, encourage, restrict or reinforce – without that activity no shaping can take place. The mathematics teacher is the one person charged, by society, with the responsibility for that 'signalling activity' in a pupil's mathematical education, but the teacher is not the only significant person in the pupil's mathematical environment. In the particular situation of the mathematics classroom, there are the other social constituents – the other pupils, the peers. They all signal their feedback, and they all play their part in helping each other shape their mathematical ideas.

5. THE TEACHING PROCESS AND MATHEMATICAL VALUES

What then can we say about the teaching process in relation to the mathematical values described earlier? What significance do we give them? What significance should we give them? How does this process develop mathematical values? How should it develop them?

I intend to explore these issues further by focussing on three particular aspects of the teaching process. The first aspect will concern the *asymmetrical* nature of the didactical situation, which is what gives the teaching process its *dynamism*. I call this aspect asymmetrical because it describes the power imbalance which drives the interaction. It is clearly likely to be an important aspect in an essentially interpersonal process, but I would argue that it is even more important when we are considering values because values don't exist in any tangible or objectified form – they are 'borne' by people – teachers and pupils are value-carriers. Thus any consideration of values in the teaching process must take into account the people concerned and the power relationship between them.

The second aspect, which I shall call the *intentional* aspect, will be concerned with the goals of mathematics education. If we want to make values a more explicit part of a pupil's mathematical education then clearly we need to focus on the aims and intent. In this case how should goals reflect mathematical values? How should they direct the teacher's choices and decisions during the teaching process? To what ends should teachers be helping the pupils with 'shaping' their ideas?

The third aspect of the didactical situation will concern its *ideational* quality – the fact that values are concerned with mathematical *ideas*. This fact focuses our attention on the process of communicating and sharing mathematical values, and requires us to generate appropriate activities for the mathematics classroom. Mathematics is, like any part of our knowledge, socially constructed, and we need to create ways of incorporating the social construction of this knowledge into the teaching process, if we want the pupil's mathematical education to take mathematical values seriously.

Let me therefore consider these aspects in relation to the mathematical teaching process.

6. THE ASYMMETRICAL ASPECT OF THE MATHEMATICAL TEACHING PROCESS

This aspect refers us to the essential power imbalance which exists in the classroom. The mathematics teacher is appointed by the society, on the basis of her superior knowledge and training and has both formal authority and sapiential authority by virtue of her position. The pupils *must* be at school, they have little choice, and they are clearly in a relatively powerless position, both at school and in the classroom. It is this asymmetry which drives the teaching process.

In relation to values, the teacher is both a value-carrier, and a value-mediator. She is a value-carrier because of her own mathematical education through which she will have developed certain attitudes, beliefs and values, and she is a value-mediator because of her role in helping the pupils share and develop their mathematical ideas.

As a value-carrier the teacher is therefore acting as some sort of model. The values she has assimilated are likely to affect all aspects of her work – for example, her choice of topics, her emphasis towards certain kinds of mathematical activities, her preferred teaching style and the support or criticism which she gives to her pupils' efforts in class. We know from the 'attitude' research literature that teachers' attitudes generally exert a controlling influence on their pupils' attitudes, hence the in-service attempts to make teachers aware of this effect (see for example, McLeod, 1987).

The educational problem occurs when the mathematics teacher, acting as a value-model, only represents these values to her pupils implicitly. It is likely that power is greater when the knowledge is implicit, and this is why I called this area, early on in the chapter, part of the 'hidden' curriculum. It exerts a powerful influence on pupils, but often without them, or their teacher, being aware of it.

The need is for the teacher to take the initiative and to find ways to make values more explicit in the classroom. In general this seems to require two kinds of approaches. The first is when there are spontaneously arising incidents in the everyday classroom context. There are always moments where the pupil's mathematical value system is exposed and you can hear this from their statements and questions – "I have the answer, so why do you want to see my method?", "I think Jane is right – she always does better than me", "Don't confuse me with another way!". At such moments the possibility exists for the teacher to educate about mathematical values, but will this opportunity be lost in the concern for the immediate mathematical task and its solution? The teacher, at such moments, does have the chance to clarify the values as well as being able to help the pupils articulate their own values. For example, with the pupil who prefers Jane's method there is a good opportunity to educate both her and the class about the rationalism value. Equally the value of openness could be exposed by taking up in a general discussion the question of "Why do you want to see my method?"

There is however still a difficulty in that the teacher may just be clarifying her own values, and the pupils may still be just learning these values being modelled by the mathematics teacher. She is still acting as a value-carrier in this situation. So how can the shift be made to the role of value-mediator?

I believe that this is where we need a second approach, and where mathematics teachers need to initiate some kinds of mathematical activities which focus the pupils' *direct* attention on mathematical values:

- values awareness activities,
- values clarification activities,
- values conflict activities,
- values criticism activities.

As one example (more will be given in the section on ideational aspects) it is very possible for a teacher to choose three pupils' methods of solving a particular problem, to have the pupils write these on the board, and to encourage a discussion on which method is better and why. Indeed a general strategy would be always to take time, following some problem-solving activity, to discuss what happened, why, and what has been learnt which can help next time. This is not a natural thing for pupils to do, nor is it at present a typical thing for teachers to do, but I firmly believe that class or group *reflection* following an activity offfers an excellent context for dealing specifically with values.

But how should the teacher deal with these reflections and discussions? Can the teacher be neutral in respect of values? Experience from the humanities subjects suggests that this is not feasible (see Metcalf, 1971). I believe that the more appropriate position for the teacher to take, bearing in mind the asymmetrical aspect of the situation, is that of facilitator. Her task should be to expose criteria and values, to make her pupils more aware of the range of values, to contrast different views and to show that there are other aspects of mathematics which are important to consider besides merely technical aspects. (Frankenstein, 1989 illustrates well the educator's role in injecting values education into mathematics teaching.)

Nevertheless the issue remains of how teachers, in their dominant role, can best educate their pupils about mathematical values when they themselves have their own feelings about these values. The teacher is clearly in a powerful position but with that power comes a responsibility to act in the pupils' and in society's best interests. This aspect is clearly highly important and sensitive when the pupil's home cultural background differs in some way from that represented by the teacher and the school.

7. THE INTENTIONAL ASPECT OF THE MATHEMATICAL TEACHING PROCESS

As we have seen above, allied to the asymmetry in the power relationship between teacher and pupils, which provides the motive force for the didactical process, is the fact that the teacher *intends* something particular to happen with the pupils. Teaching is clearly not a random activity, although sometimes it may not be clear precisely what the teacher's intentions are. But one cannot begin to teach without intentions. So how does this intentional aspect relate to mathematical values? How does the mathematics teacher contextualise in *her* goals the kinds of values intentions referred to in Section 2?

If indeed it is true that teachers do not in general have specific intentions about making mathematical values explicit in their pupil's education, it is likely that the majority of this aspect of their education (if we can call it that) will occur implicitly. However, as I have said earlier, just because values are implicit in much of the curriculum, it doesn't mean that mathematical values are not learnt. They are learnt, and many adults are very willing to express their views regarding the values they learnt about in mathematics classrooms (as we have seen, for example, in Buxton, 1981).

Moreover, my conjecture would be that for the majority of adults, their mathematical values relate primarily to the objectism, control and mystery clusters, and Buxton's respondents support that conjecture. For most people mathematics is a collection of objective facts, indisputably universal and generally unrelated to each other. There may be the occasional memory of somebody's Theorem (usually Pythagoras') but that in no way makes sense in terms of rational theorising. The objectism values are clear.

Equally, mathematics to most people represents secure knowledge: "You know where you are with Maths, that's why I liked it" is a typical comment. Knowing that mathematics helped produce the Pyramids, helped win a war or two, and helped to land a man on the moon confirm for most people its *control* and predictive value. The ideas of alternativism, change, and progress are not so obvious nor are they so acceptable. For many adults, it is very disconcerting to be faced with other 'facts' like, for example, that there are 500 different counting systems in Papua New Guinea, that the Navajos have a very different conception of space from us, and that there are not always 180 degrees in a triangle (two meridians and the equator). The first two usually are explained away as being 'primitive', while the last is merely a 'trick' answer!

Certainly the value of openness is not commonly held. For most adults, mathematics is a very mysterious phenomenon indeed, which "only a real mathematician can understand". The fact that it is dehumanised knowledge also contributes to this image for them. Mathematicians are considered strange, 'other' beings, often very absent-minded and illogical in their normal lives, if they have any normal lives! The basic facts like $2 + 2 = 4$ are clearly not felt to be mysterious, but for many people those don't constitute the 'real' mathematics. For them, real mathematics begins when they don't understand it and mystery takes over! (Usually this feeling begins with the studies of negative numbers, fractions, algebra, or infinity.)

So, I contend that because of the emphases given in the canonical mathematics curriculum, and because of the *implicit* learning of values, people leave school with a set of mathematical values which are unbalanced. It therefore seems to me essential to expose this state of affairs, and to consider how we can change both the intended mathematics curriculum and the teaching process in order to make values learning more explicit, and thus to be able to give more consideration to the other value clusters of mathematics. The teacher should in my view have the intention to expose her pupils to the full range of mathematical values, as described earlier.

Furthermore, it seems to me, personally, to be crucial that in the highly technological societies of Europe, which are also increasingly participant democracies, the need is more and more for mathematics teachers to offer the values of rationalism, of progress, and of openness. If we want our education in general to develop critically aware and democratically conscious young people then mathematics education must play its particular part, by emphasising explicitly those values which can contribute to that education.

8. THE IDEATIONAL ASPECT OF THE MATHEMATICAL TEACHING PROCESS

If mathematics teachers have the power to do this, and also the intention, how can they make it happen? This question relates to the ideational aspect of the teaching process – the fact that the process is about the teaching and learning of ideas. Ways have to be created to enable mathematical values to be communicated and shared.

As I have already indicated, I feel that one way this can be done is to give more opportunity for discussion in lessons. Strangely, I have often had the experience of talking about this with teachers who, while they themselves are happy to discuss mathematical ideas, seem to feel that there is nothing which could be discussed in class: "But what should they discuss?" Again, as I said earlier, different methods of solving a problem can be an easy stimulus for such a discussion. One objective for the mathematics teacher here might concern how she could arrange for her pupils to share the knowledge which *she* has concerning the criteria for the acceptability of different methods. At the very least, this *intention* will make the teacher become more aware of her own values. I realise that mathematics classrooms do not usually contain much discussion, so the teachers need help, supportive materials and different strategies if they are going to work on this area. I firmly believe it is one of the most important areas of strategy which mathematics teachers need to develop.

Another aspect of teaching which could further this aspect of the teaching process is that of small group work. The social arena of the whole class situation is not a very easy one for developing discussions and it would be much more productive in my view to have more small group work. Suitable activities have been devised but if we just stay with the 'different solutions to the same problem' example it would be an excellent activity for each small group of pupils to compare problem-solving methods, to describe how they differ, then to decide which methods are better than others, and to clarify what their criteria are for their decisions. Into these interactions it would be relatively easy for the teacher to inject ideas about rationalism and openness, for example.

Equally, small groups are excellent contexts for generating multiple approaches to problems and to other tasks. These clearly relate to the values of alternativism and progress, but once again these will lie dormant, though they may be learnt implicitly of course, until exposed by the teacher's intervention and reflected upon. In fact I believe that the absence of small group working in most mathematics classrooms undoubtedly contributes to the general lack of understanding and belief in the value of alternativism and progress in mathematics.

A different approach which can facilitate the communication and sharing of values is to vary the focus of the mathematical activity. Two particular examples of this are projects, and investigations. *Projects*, are I believe excellent activities for the consideration of relationships between mathematics and society, and of the values concerned with those relationships. I would distinguish between three timescales. First, there are historical projects which can concern the social history of mathematical ideas and which can demonstrate the growth of ideas and the development of values. Second, there are projects concerning present-day society, which can demonstrate the roles mathematical ideas and values play in the present society. Third, there are projects on society in the future in which potential solutions to environmental problems, for example, can be explored.

In each case the projects can be done with the whole-class, in small groups, or by individual pupils. All three group-sizes have their different merits but I think individually-based projects have much potential. The pupil has to learn to use a variety of material resources and has to develop a critical faculty. Also the possibility of individuals choosing topics which particularly interest them has the potential of making any values-education through those projects more meaningful and significant for the pupil. (Bolt and Hobbs, 1989, contains many examples of mathematical projects suitable for the classroom.)

Then there are mathematical *investigations*, an activity where the pupil does creative, exploratory work on a mathematical idea. In a good investigation the pupil must define some terms, symbolise some variables, explore and describe relationships, and demonstrate the truth or otherwise of those relationships. A great deal of creative mathematical activity can be done and the pupil thereby experiences the decision-making, the criteria and the values, of the practising mathematician. As with projects, mathematical investigations can be done in different sized groups but again I would maintain that there is much to be gained by pupils working individually. Also investigations need to be written about, as do projects, and that more-reflective part of the activity once again gives scope for discussion of values. (Mottershead, 1985, gives some excellent examples of mathematical investigations suitable for the classroom.)

Thus by using discussions, small group work, mathematical projects and investigations, the teacher can create a variety of learning contexts in which the education of mathematical values can happen. Note that I say 'can' because it still demands the intention on the part of the teacher to focus her pupils' attention on mathematical values, and it also requires an appreciation of the asymmetrical aspect of the teaching process to ensure that the pupils don't just assimilate the teacher's personal views.

CONCLUSION

In this chapter, then, I hope I have developed enough ideas to demonstrate that further analysis of mathematical values is necessary, that values education is important and desirable within the mathematics curriculum, and that it is very possible for teachers to become more professional about this aspect of mathematics education.

Perhaps the most significant point in this chapter is my belief that much values-learning does in fact take place in mathematics classrooms, but *implicitly*. As a result, it is *not* producing desirable feelings and attitudes, and we should therefore take steps as soon as possible to develop ways, like those described above, to make this area of mathematics education more explicit and more meaningful. Only then will it *contribute* positively to the pupil's mathematical education rather than detract negatively from it, as I believe it does at present.

REFERENCES

d'Ambrosio, U.: 1985, 'Ethnomathematics and its Place in the History and Pedagogy of Mathematics', *For the Learning of Mathematics* **5**,1, 44–48.

Apple, M.W.: 1980, 'The Other Side of the Hidden Curriculum: Correspondence Theories and the Labor Process', *Journal of Education* **162**, 47–66.

Bishop, A.J.: 1988a, *Mathematical Enculturation: A Cultural Perspective on Mathematics Education*, Kluwer Acad. Publ., Dordrecht.

Bishop, A.J.: 1988b, 'Mathematics Education in its Cultural Context', *Educational Studies in Mathematics* **192**, 179–191.

Bishop, A.J.: 1990, 'Western Mathematics: The Secret Weapon of Cultural Imperialism', *Race and Class* **32**(2), 51–65.

Bolt, B. and Hobbs, D.: 1989, *101 Mathematical Projects*, Cambridge University Press, Cambridge.

Buxton, L.: 1981, *Do You Panic about Maths: Coping with Maths Anxiety*, Heinemann Educational, London.

Dijksterhuis, E.J.: 1986, *The Mechanization of the World Picture – Pythagoras to Newton*, Princeton University Press, New Jersey.

Frankenstein, M.: 1989, *Relearning Mathematics: A Different Third R – Radical Maths*, Vol. 1. Free Association Books, London.

Gardner, P.L.: 1977, *Logical Connectives in Science*, Monash University, Faculty of Education, Melbourne, Australia.

Gerdes, P.: 1986, 'On Culture, Geometrical Thinking and Mathematics Education', *Educational Studies in Mathematics* **19**,2, 137–162.

Gerdes, P.: 1988, 'A Widespread Decorative Motif and the Pythagorean Theorem', *For the Learning of Mathematics* **8**,1, 35–39.

Joseph, G.G.: 1987, 'Foundations of Eurocentrism in Mathematics', *Race and Class* **XXVII**,3, 13–28.

Kline, M.: 1972, *Mathematics in Western Culture*, Pelican, London.

Lawler, R.: 1982, *Sacred Geometry*, Thames and Hudson, London.

Levidow, L. 1988, 'Non-Western Science, Past and Present', *Science as Culture*, Free Association Books, London, pp. 101–117.

McLeod, D.: 1987, 'Beliefs, Attitudes and Emotions: The Affective Factors in Mathematics Learning', in J.C. Bergeron, N. Herscovics, and C. Kieran (eds.), *Proceedings of P.M.E.XI conference*, Montreal, Canada, pp. 170–180.

Mellin-Olsen, S.: 1987, *The Politics of Mathematics Education*, Kluwer Acad. Publ.,

Dordrecht.
Metcalf, L.E.: 1971, *Values Education*, 41st Yearbook of the National Council for the Social Studies, N.C.S.s., Washington D.C.
Mottershead, L.: 1985, *Investigations in Mathematics*, Blackwell, Oxford.
Pinxten, R., van Dooren, I., and Harvey, F.: 1983, *The Anthropology of Space*, University of Pennsylvania Press, Philadelphia.
Ronan, C.A.: 1981, *The Shorter Science and Civilisation in China*, Vol. 2, Cambridge University Press.
Swadener, M. and Soedjadi, R.: 1988, 'Values, Mathematics Education, and the Task of Developing Pupils' Personalities: An Indonesian Perspective', *Educational Studies in Mathematics* **19**,2, 193–208.
Van der Waerden, B.L.: 1983, *Geometry and Algebra in Ancient Civilizations*, Springer, Berlin.
White, L.A.: 1959, *The Evolution of Culture*, McGraw-Hill, New York.
Wilder, R.L.: 1978, *Evolution of Mathematical Concepts*, Open University Press. (First published by J. Wiley, 1968.)
Wilson, B.J.: 1986, 'Values in Mathematics Education', in P. Tomlinson and M. Quinton, *Values Across the Curriculum*, The Falmer Press, Lewes, pp. 94–108.
Zaslavsky, C.: 1973, *Africa Counts*, Prindle, Weber and Schmidt, Boston, Mass.

INDEX OF NAMES

INDEX OF SUBJECTS

Mathematics Education Library

Managing Editor: A.J. Bishop, Cambridge, U.K.

1. H. Freudenthal: *Didactical Phenomenology of Mathematical Structures*. 1983
   ISBN 90-277-1535-1; Pb 90-277-2261-7
2. B. Christiansen, A. G. Howson and M. Otte (eds.): *Perspectives on Mathematics Education*. Papers submitted by Members of the Bacomet Group. 1986. ISBN 90-277-1929-2; Pb 90-277-2118-1
3. A. Treffers: *Three Dimensions*. A Model of Goal and Theory Description in Mathematics Instruction – The Wiskobas Project. 1987 ISBN 90-277-2165-3
4. S. Mellin-Olsen: *The Politics of Mathematics Education*. 1987
   ISBN 90-277-2350-8
5. E. Fischbein: *Intuition in Science and Mathematics*. An Educational Approach. 1987 ISBN 90-277-2506-3
6. A.J. Bishop: *Mathematical Enculturation*. A Cultural Perspective on Mathematics Education. 1988
   ISBN 90-277-2646-9; Pb (1991) 0-7923-1270-8
7. E. von Glasersfeld (ed.): *Radical Constructivism in Mathematics Education*. 1991 ISBN 0-7923-1257-0
8. L. Streefland: *Fractions in Realistic Mathematics Education*. A Paradigm of Developmental Research. 1991 ISBN 0-7923-1282-1
9. H. Freudenthal: *Revisiting Mathematics Education*. China Lectures. 1991
   ISBN 0-7923-1299-6
10. A.J. Bishop, S. Mellin-Olsen and J. van Dormolen (eds.): *Mathematical Knowledge: Its Growth Through Teaching*. 1991 ISBN 0-7923-1344-5
11. D. Tall (ed.): *Advanced Mathematical Thinking*. 1991 ISBN 0-7923-1456-5
12. R. Kapadia and M. Borovcnik (eds.): *Chance Encounters: Probability in Education*. 1991 ISBN 0-7923-1474-3

KLUWER ACADEMIC PUBLISHERS – DORDRECHT / BOSTON / LONDON